Poultry and Egg Production

by Harry Reynolds Lewis

with an introduction by Jackson Chambers

This work contains material that was originally published in 1916.

This publication is within the Public Domain.

This edition is reprinted for educational purposes and in accordance with all applicable Federal Laws.

Introduction Copyright 2017 by Jackson Chambers

IMPORTANT NOTE & DISCLAIMER

IMPORTANT NOTE :
As with all reprinted books of this age that are intended to perfectly reproduce the original edition, considerable pains and effort had to be undertaken to correct fading and sometimes outright damage to existing proofs of this title. At times, this task can be quite monumental, requiring an almost total rebuilding of some pages from digital proofs of multiple copies. Despite this, imperfections still sometimes exist in the final proof and may detract slightly from the visual appearance of the text.

DISCLAIMER :
Due to the age of this book, some methods or practices may have been deemed unsafe or unacceptable in the interim years. In utilizing the information herein, you do so at your own risk. We republish antiquarian books with no judgment or revisionism, solely for their historical and cultural importance, and for educational purposes.

Self Reliance Books

Get more historic titles on animal and stock breeding, gardening and old fashioned skills by visiting us at:

http://selfreliancebooks.blogspot.com/

INTRODUCTION

I am very pleased to present to you another essential poultry title – **Poultry and Egg Production**. It was written by Professor Harry Reynolds Lewis in 1916. Professor Lewis was the *Director of Poultry Husbandry* at the *State Agricultural College,* New Brunswick, New Jersey.

With the rise in popularity of back-yard farming we love to bring you as many useful poultry-related titles as we can. People are turning to raising their own fruits, veggies, meat and eggs to avoid the disease, sub-par nutrition, and escalating costs of buying chemical-laden and feed-lot-raised commercial food.

By growing our own food, raising our own meat and eggs, we can avoid the pitfalls of mass-produced food and bring to our dinner tables healthy, natural, nutritious food for our families. And once you've tasted a home-raised egg – with a yolk that is brilliant orange, not and insipid yellow – you will never want to buy a box of eggs at the store ever again!

Lewis expertly walks us through the vital topics of principles and practices of poultry and egg production, prevention and precautions against disease, hatching and rearing, feeding, sanitation, and many more essentials.

His book covers the basics – the principles and fundamentals – of poultry and egg production, and is a great place to start for the novice poultry breeder, or those just thinking about taking the plunge into poultry farming.

Jackson Chambers

State of Jefferson, December 2017

PREPARATORY NOTE.

In view of the increasing value of the Poultry and Egg Industry in the State of New Jersey, and the importance of proved directions and instructions for success in the business, the Executive Committee of the State Board of Agriculture requested Prof. Harry R. Lewis, Director of Poultry Husbandry, State Agricultural College, New Brunswick, N. J., to prepare a comprehensive bulletin on this general subject for circulation among the farmers and poultry producers of the State. Prof. Lewis has kindly prepared this bulletin in answer to the request of the Executive Committee and the Committee believe that if its directions are followed the annual returns from our poultry will be greatly increased.

JOS. S. FRELINGHUYSEN,
President.

FRANKLIN DYE,
Secretary.

Trenton, N. J., August 2, 1916.

PLATE NO. 1.—A typical poultry farm lay out, which is run as a side line to general farming. This scene is from Cape May County.

POULTRY AND EGG PRODUCTION.

New Jersey Conditions. New Jersey has, within the last few years, come to be known as one of the greatest egg producing sections in the world. Her conditions are somewhat different from the majority of places, because the best markets in the world are at her very door. The problem of the poultryman becomes one of efficient production while the problems of distribution are of minor importance. There are, within New Jersey, three distinct types of poultry keepers. First, the so-called commercial poultryman, who is usually a specialist, devoting his entire time and entire farm to the production of poultry and eggs. The second type of poultryman may be termed the farm flock man, for poultry is kept in small flocks as a side line to general farming. Lastly, we come to the suburban or urban poultryman, who keeps a few chickens in his back yard to supply his own table. The problem of each of these three types is materially different. A brief discussion of problems as they exist will not be out of place. The Poultry Department at New Brunswick, for the past three years, has been securing census data pertaining to the distribution of poultry. To date, they have located approximately one thousand farms that winter over one thousand birds each. All of these farms can be called specialized commercial types. Of these one thousand farms, over two hundred winter over two thousand birds each. There are two farms in the state, one which winters twelve thousand, and the other eighteen thousand. In comparison with other states in the Union, New Jersey is essentially the state of large poultry farms. On the majority of these commercial plants, the production of white eggs for market is the primary consideration. Running secondary to this is the growing of broilers and the rearing of cockerels and pullets for sale. The problem of the commercial poultryman is one which entails a complete knowledge of the business and a careful organization of the work, in order that the enterprise should be financially successful.

There are in New Jersey 35,000 farms. It is doubtless a safe estimate to say that at least 95% of these farms maintain a poultry flock ranging in number from ten to four or five hundred birds, which figure gives some indication of the immensity of New Jersey's poultry industry. (See plate No. 1.) A great many of the farm flocks are run as a rather neglected side line to other farming operations, the idea being to let them get a great deal of their existence by ranging, with the expectation that enough poultry and eggs will be supplied for the home table, and that in addition, a limited supply will be available through wholesale channels. A great many of the farm flocks are poorly managed, resulting in unsatisfactory returns. With more appreciation of the possible financial revenue from the farm flock, greater attention is given to the better methods of production.

The United States census gives for the state of New Jersey over two million head of poultry, valued at over $2,000,000.00. This figure simply covers the poultry kept on farms and does not include, in any sense of the word, small flocks kept by the suburban and urban population. To determine approximately the number of birds kept in the state, a census has been taken of a number of urban centers, the average results found being exemplified in the city of New Brunswick, maintaining a population of twenty-seven thousand. A house-to-house and yard-to-yard canvass showed the presence of fourteen thousand adult birds being kept within the thickly populated city. This means practically one bird to every two people in the population. Figures very similiar to these were found to exist in other cities of both larger and smaller population.

The problem of the suburban poultry keepers is primarily to keep but a few birds, simply enough to supply the home table with fresh eggs and poultry meat. It is the ambition of many of these poultrymen to some time own a poultry flock of their own. As a class, they are very ambitious, looking for the latest information, and taking as much or more interest in their small flock than is taken by the average commercial poultry farmer. Taken as a whole, probably the best combination for the average poultryman is the production of eggs, as the main line of effort, combining with this the production of broilers in the early spring in so far at least as the sale of surplus cockerels will admit. Pure bred birds of some recognized standard variety should be kept, as this gives a good opportunity for advertising and offering for sale hatching eggs in season, and if range conditions are sufficient and of suitable nature, it will often be a very desirable practice to grow a limited number of cockerels to maturity which can be sold for breeding purposes. Additional pullets other than required for that particular farm, for which there is always an excellent demand in the fall at very remunerative prices can also be grown. As previously mentioned, New Jersey is especially fortunate in being located in the center of the most thickly populated district in the United States, sections which have the best markets in the land, sections which, in order to secure a sufficient supply of poultry and eggs, must bring them in from the western part of the country at a sacrifice of quality. The problem of the New Jersey poultryman becomes one of efficient management to get a profitable production, as the quality will not be impaired in any way in marketing. The New Jersey products are always in demand in preference to the western product, at an increased selling price. If the greatest profit is to be realized from the eggs produced, the market quotations for a period of years show that at least a moderate production from October to March must be secured. It is at this season that the average farm flock, given no special attention, lays but few eggs. Beginning about April, which is the natural breeding season, the farm flocks all over the country begin to produce eggs in large quantities, which cause the price paid for the local product to drop materially. As a general statement it may be said that it is doubtful if there is little or any profit from egg production, if on account of improper care an egg yield of from 20 to 30% is not secured during the winter months.

PRINCIPLES AND PRACTICES.

It must be appreciated in the beginning that the three types of poultrymen above outlined, each have their different problems to solve, conditions of production being so different. It is fundamentally true, however, that regardless of the size of the flock or the conditions under which it must be kept, the principles governing the management are the same, whereas the practice or the application of the principles may vary in a slight degree. In other words, the principles governing poultry management are fundamental and are based upon the requirements of birds for nutrition, environment, reproduction, etc. The principles governing the management of the commercial flock, the suburban flock and the farm flock are fundamentally the same. The application of these principles will vary slightly with breed, with the size of the flock, with range conditions, with nature of product produced, with age of birds, and with numerous other factors which are local in nature. The successful production of poultry and eggs requires first a careful working knowledge of the basic principles underlying the care of the birds, together with suggestions as to the application of these principles and practices on the average farm.

It is the purpose of this bulletin to outline ten fundamental principles of poultry keeping, to discuss each in its application to poultry keeping as a whole and to offer suggestions for the application of these principles to small and large flock management.

BASIC PRINCIPLES GOVERNING POULTRY PRODUCTION.

The following are the ten basic principles which determine not only efficient production, but profitable production of poultry and poultry products:

1. A PERSONAL LIKING FOR AND A WORKING KNOWLEDGE OF THE POULTRY BUSINESS.
2. STOCK WHICH IS FUNDAMENTALLY HEALTHY AND VIGOROUS.
3. HOUSES WHICH ECONOMICALLY PROVIDE A SUITABLE ENVIRONMENT.
4. CAREFUL SANITARY PRECAUTION AS PREVENTION AGAINST DISEASE.
5. PROPERLY BALANCED RATIONS AND A CORRECT FEEDING PRACTICE.
6. A SYSTEMATIC EFFORT TO BREED FOR VIGOR, EGG PRODUCTION AND BODY CONFORMATION.
7. THE USE OF PROPER METHODS IN HATCHING AND REARING.
8. INSURING A HIGH QUALITY IN THE MARKET PRODUCTS PRODUCED.
9. SPECIAL CARE AND ATTENTION TO THE MANNER OF MARKETING.
10. BUSINESS MANAGEMENT IN ALL BRANCHES OF THE WORK.

The following discussion of these factors will deal especially with New Jersey conditions.

A PERSONAL LIKING FOR AND A WORKING KNOWLEDGE OF THE POULTRY BUSINESS.

The poultry business is no different than any other line of manufacturing, mercantile, or commercial enterprise. The man and the brains at the head of the business are really responsible for the success or failure of the enterprise. The poultry business, more than most other lines of agriculture, is essentially a business of small details. The poultryman to succeed must then be capable of close application to details. A lack of this one quality more than any other one will result in a failure. The poultryman must be a good student and a good observer. Much of the knowledge relative to birds can be secured by becoming intimately associated with them. A careful and conscientious application to the work is very requisite. The successful poultry keeper of today must be willing to try out new appliances and ideas, must be a good mixer in a similar business and must be able to profit by the experience of others. In studying the birds, their habits, etc., individuals should be made as much as possible a unit of study. Recent findings in the breeding and feeding of fowls indicate that the individual must be used, if continued improvement is to be secured. It is a fundamental principle of any business that success can only be continually secured where the one operating same has a general aptitude and liking for the occupation. With this natural interest for the work and with a good practical training, backed by as much theoretical knowledge as can be secured, the commercial production of market poultry and eggs, will bring a good living to the poultryman and satisfactory returns on the money invested. Exceptionally high or phenomenal prices cannot be secured nor should they be expected. A yearly net profit per bird of from $1.00 to $2.00 is on the average a safe estimate. The actual returns will depend largely upon market conditions and upon the quality of the product produced. Where greater profits are realized than those above enumerated, the increase will usually come from the sale of eggs for hatching and from the sale of adult birds for breeding or laying purposes.

To the amateur contemplating starting in the business it should be clearly stated that the start should not be made unless sufficient capital is available and unless the one contemplating investing his money has a satisfactory knowledge of what is ahead of him. This knowledge should be secured in three ways, first, by the reading of poultry papers, periodicals and reliable text books, secondly, by the working for at least one year on some moderate sized poultry farm, primarily for experience, and, lastly, by the pursuing of some winter course in poultry husbandry at the state agricultural colleges. Information secured in these three ways will be complete and sufficient to enable a proper beginning.

THE FOUNDATION STOCK MUST BE FUNDAMENTALLY VIGOROUS AND HEALTHY.

Constitutional or inherited vigor is, above everything else, necessary in order that the poultry flock shall succeed continuously. It is the experience of the author that where the stock is weak, poor hatches result, high mortality is common, low egg production and sick birds are the rule. On the other hand, where poultry farms have made marked success, it can be traced in almost every instance to sturdy, rugged, healthy, vigorous stock. In discussing the stock, there are a number of things to consider. Foremost is the determination of the breed best adapted to produce the type of product required. The birds themselves have frequently been likened to machines, which are required to transform the raw product, feed, into a finished product, eggs and meat. Without a well built machine, this transformation cannot be carried on at a profit. There are many different types of birds, each designed for the production of a different type of product. The determination of the breed should be one of the first questions of solution. Three general classes of birds exist, the so-called egg or light breeds, most of which are of Mediterranean origin, of which the Leghorn is the typical example. (See plate No. 2.) These birds are kept in large numbers on commercial farms where the production of white eggs for the wholesale market is the primary object. It is an established fact that these white shelled eggs bring an increased amount over brown eggs of from 3 to 15 cents, the average being about 5 cents per dozen during the entire year. These light egg breeds stand the confinement well and admit of herding together in large flocks. They are essentially active and are close feathered. Their rather large, fleshy head parts require careful protection against freezing during severe weather in the winter. The second type of bird may be called the general purpose or dual purpose fowl, examples of which are the Plymouth Rocks, the Wyandottes, the Rhode Island Reds, and the Orpingtons. (See plates 2 and 3.) These breeds are characterized by their ability to lay a goodly number of brown shelled eggs, and in addition their large size and heavy weight make them capable of bringing in considerable revenue when sold for market purposes. It is this type of bird which is kept on some of the large commercial farms, but they more especially comprise the popular type on the farm and in the suburban communities. Being largely dual purposed, they are the most desirable type for supplying home demands. The third type of fowl is known as the meat breeds, of which the Brahma and Langshan are examples. These are the largest birds kept and are noted for their fine quality of meat. Their ability to produce eggs, however, is limited. These large types are kept in flocks where market poultry products are the primary object, and especially where capons and roasters constitute the main marketable product. Recognizing these three distinct types, the breed should be selected which personally appeals to the poultryman, and which produces the best type or combination of products which is especially adapted to his particular object or markets.

Standard Bred Poultry the Best. We hear a great deal these days about utility vs. fancy which, analyzed, means, the production of eggs and meat vs. shape and plumage pattern. This agitation about the so-called utility possibilities has doubtless been brought about by the results secured at egg laying contests throughout the country, and also by some phenomenal records which have been made by some cross bred birds. As a general statement, it may be said that it is a mistake to attempt to produce eggs or poultry continuously from cross bred stock. Considered over a period of years, birds which are continually cross bred produce no characteristics which are supplied by those possessed by pure bred individuals. Cross bred birds show no reliability in breeding where egg production is the primary object. A larger egg production can be secured from typical egg breeds than can be secured from crosses of the egg and dual purpose type. Where meat production is the object, larger returns both in quantity and quality can be secured where the pure bred meat breeds are used. The use of standard bred birds, of a well-established variety, results in a uniform flock, both as to general appearance, size, shape and plumage pattern, which make them much more attractive and interesting. The eggs produced by pure bred individuals are uniform in size, shape and color, and bring a more uniform price on the market than mixed types. They cost no more to keep than a mongrel or a bird of mixed breed, for they consume no more food nor do they require any more labor to care for them. Where pure bred birds are kept, often considerable revenue can be secured from stock and eggs sold for breeding purposes. The additional revenue so secured is almost all clear profit. These factors should be given careful consideration before an attempt is made to produce eggs at a profit from a flock which has been promiscuously cross bred.

The term utility has recently come to be quite generally misused—cull birds and scrubs resembling in no degree the standard bred birds, have been termed utility, the thought being portrayed that they were remarkable for egg production or for some other utility character. This, however, is often not the case, as the poultryman owning these birds keeps no complete records of breeding or of production. The proper and the only successful feature lies in combining utility qualities of good production, vigor, amount and quality of flesh with standard bred varieties. The best solution of the problem will be for the poultryman to select a breed which, owing to its natural adaptation, best meets his particular conditions. After choosing the breed, secure a strain of this breed which is of known reputation, in so far as stamina, egg production, and body conformation is concerned, and then having secured this desirable strain as foundation stock, systematically breed year after year to retain these qualities, and, if possible, increase them, working always with the expectation of combining in one individual the ideal standard shape, color and general appearance required for that variety, together with high egg production and vigor. Do not let the term utility, unless it is backed up with records and known performances, mislead one into securing scrub birds or culls.

Only healthy birds should be kept on the plant. The presence of disease of even a mild form shows an inherited weakness.

Roup, chicken pox, canker, diarrhœa and digestive disorders of any kind, if present in birds properly cared for, indicate that the birds so affected are inferior physically. It should be the general practice to get rid of birds so affected to the best possible advantage, never keeping them over for a second year's lay, and, above all things, never use them for reproduction purposes.

THE POULTRY HOUSE SHOULD BE ECONOMICALLY CONSTRUCTED AND SO DESIGNED AS TO PROVIDE A SUITABLE ENVIRONMENT.

The design of the poultry house and of its construction determines in large measure the environmental conditions which will surround the bird. The modern poultry house must provide all the features necessary to create ideal conditions, for no matter how well a flock of birds may be bred, or how well they may be fed, if the poultry house is damp, drafty, cool, and poorly lighted, disease and a decrease in vigor will surely result.

Dryness Is Fundamental in the Poultry House. Moisture breeds disease. Moisture in a poultry house may be caused in a number of different ways, first, the atmospheric or air moisture, which is usually caused by imperfect ventilation. The presence of this moisture can be determined by drops or beads of water which will be seen on the rafters and walls of the house during damp weather. Changes in temperature cause this atmospheric moisture to become condensed in the form of drops of water. A second type of moisture is that which is exhaled by the birds in breathing. Unless carried off by proper ventilation, this exhaled moisture not only makes the house damp, but also results in filling the air with impurities. The third type of moisture is that which comes from the soil. This is especially prevalent if the soil is heavy, of a clay nature, and low. When the house is located on such spots, the danger of soil moisture must be counteracted by proper drainage and the construction of suitable floors. Sometimes moisture in the house may be due to surface water running in. This is apt to be true in the spring of the year, when the ground is frozen. On this account the house should always be located on a hill, and never in a low hollow area. Perfectly dry floors and dry, pure air are the first requisites.

Plenty of Sunlight Needed. The poultry house which is flooded with sunlight is much more sanitary and healthful and lends to the contentment of the flock. (See plate No. 5.) Direct rays of sunlight are the best destroyers of disease germs known. In order to get a free access of sunlight a house should be placed so that it faces in a southerly direction. The design should provide a front from seven to nine feet in length, and a house of such depth that at some time during the day the sunlight will strike every part of the floor. The openings in the front which admit of sunlight can be covered with part glass and part muslin, muslin being used in stormy weather to keep out the dampness, and at the same time to admit of free circulation of air.

Importance of Thorough Ventilation. Ventilation in the poultry house is especially important during the winter, when it is essential to keep a large number of birds closely confined to their quarters. Fowls are very active and they possess a high body temperature which results in the exhaling of a large amount of impure air which contains carbon dioxide, a dangerous poison. The health of the flock demands that this poison be removed and the impure air be replaced with pure oxygen laden air. This impure air breathed out by the birds is also heavily laden with moisture. Complete ventilation liberates this moisture and keeps the house dry. In providing ventilation, excessive drafts should be avoided. There is probably no one thing which will bring on roupy conditions more than a drafty house. Drafts can be avoided by having the openings in the front two or three feet above the ground, and in the case of a laying house by breaking it up in twenty or thirty-foot sections with cross partitions. Proper ventilation can be provided in a number of ways. A liberal use of muslin in the front is doubtless the first requisite. If the glass and muslin windows are left open much of the time, the atmospheric conditions within the house will be as pure as those out of doors. During wet, stormy weather, especially in the winter, the muslin curtains in the front can be closed, and if a light weight open cloth is used, the air passes through it readily. Slatted shutters resembling blinds are sometimes recommended in the place of the muslin. These should be hinged so that they can be opened when needed, otherwise they shut out considerable sunlight. During the summer, low shed-roof houses with black tar paper covered roofs are apt to be exceptionally hot. Ventilation under the back eave so arranged that the northerly winds can enter the house and circulate between the studding on the back wall and the roof, has by test, proven very desirable. Cupola ventilators are not necessary and are of questionable advantage in the poultry house.

Give the Birds Plenty of Floor Space. In planning the laying house, where the birds are to be confined during the winter, the number of birds which the house will accommodate should be determined largely by the floor space. They must have all the floor space which they need for exercise, and not be so crowded that it will be impossible to maintain satisfactory sanitary conditions. Since the house is built for the birds and its capacity is determined by the floor space, all interior fixtures, such as nests, hoppers, fountains, etc., should be elevated above the floor, giving the entire floor space to the flock. A safe rule to follow is to allow Leghorns or the light active breeds approximately four square feet of floor space per bird. In the case of the heavier breeds, such as the Plymouth Rocks, it is doubtful if they can get along satisfactory with any less than five square feet. As the floor space per bird is decreased, or as the birds are crowded, an increasing proportion of time and money must be spent in maintaining sanitary conditions, or disastrous results will follow such intensification.

Birds Must Be Protected From an Excessive Cold Temperature. It is a long established fact that birds will stand a remarkably low temperature, providing the atmosphere is free from drafts and dry. On the other hand, a moderately low temperature of about 25 degrees, if the air is drafty and moist, will produce frozen wattles and combs. Freezing of the head parts is more common with the light Mediterranean breeds than with the heavier American or Asiatic types, for the reason that the head parts are much larger and more susceptible to freezing. During the winter the flock should be induced to take all the exercise possible, by feeding all the grain in litter. This exercise induces circulation, and by keeping the blood flowing rapidly prevents freezing. The feeding of considerable whole or of cracked corn will help to prevent disastrous results from excessive cold temperatures. In spite of all these precautions, however, the house itself must be designed and furnished, with the view of protection against cold. With this in view, the dropping boards and perches should be placed at the low back part of the house,.farthest removed from the openings in the front. If the back wall and roof over the dropping boards are double boarded, it offers a double protection. In very cold climates, a cloth curtain which can be dropped part way down in front of the perches may by intelligent use protect against excessive cold temperature. Frozen combs and wattles, while not permanently injuring the birds, put them temporarily out of condition in so far as egg production is concerned. It usually requires from two weeks to a month to get the birds so affected into normal laying condition.

The Construction of the House Must Be Rat Proof and Vermin Proof. Rats are a serious scourge on most poultry plants. The house should be so planned and built as to make it impossible for them to find permanent hiding places about it. The best protection is to construct a foundation of concrete six inches wide, and about two feet deep. After the wall is built and hard, a concrete floor about three inches thick over a drainage coat of cinders insures absolute protection against these rodents. On the range where the youngsters are reared, rats will do serious danger if the houses are built on the ground. For this reason, the colony house may be elevated about a foot from the ground. Where the poultryman desires to use a dirt floor, it is possible to lay one-half inch mesh poultry netting over the floor, nailing it securely to the sill, laying it about six inches under the ultimate top of the dirt floor. All interior construction should be built of planed or dressed lumber, the boards should be carefully matched together in order that as few hiding places as possible will be offered for mites. All interior fixtures should be portable and as simple as possible in design. This admits of easy and perfect cleaning.

Economy in Construction is a First Hand Consideration in Designing and Building the Laying House. Economy should be the first and ever present thought in the minds of the builder. Cheapness, however, at a sacrifice of efficiency is not economy. Economy in construction, combined with the greatest efficiency, can undoubtedly be secured by building a house of the shed-roof type, the back wall being from four and one-half to five feet in height and the front wall from seven to nine feet in height. The height will depend somewhat upon the depth of the house, but it must be high enough to admit the attendant caring for the birds with no inconvenience to himself. The house should approach a square in shape, for this construction is more economical, as less building material is required. The following table gives the possible combinations of such a standard unit type, all of which will be seen are constructed on the basis of a square.

Application of standard unit idea:

Size of house:	Floor space:	Capacity:
6 x 8 feet.	48 sq. ft.	10 hens
8 x 8 feet.	64 sq. ft.	20 hens
10 x 10 feet.	100 sq. ft.	25 hens
12 x 12 feet.	144 sq. ft.	40 hens
14 x 14 feet.	196 sq. ft.	50 hens
16 x 16 feet.	236 sq. ft.	60 hens
18 x 18 feet.	324 sq. ft.	80 hens
20 x 20 feet.	400 sq. ft.	100 hens

With these few words as to the principles of design, the following discussion of the New Jersey State Multiple Unit Laying House will give the details necessary to construct an efficient house possessing the features above mentioned.

The New Jersey Multiple Unit Laying House. (See plate No. 4.) The following plan of a shed-roof house 20 x 40 feet is especially suited to New Jersey poultry farms. Where it is desirable to keep larger units than a forty-foot house will accommodate, it is recommended that the length be doubled, making it 20 x 80 feet, with three cross partitions (one every 20 feet), instead of only one, as in the forty-foot house.

The following description of the above plan shows the important features:

The outside dimensions are 40 x 20 feet, sills to be 4 x 6, and to be bolted to a concrete foundation wall eight inches wide, which is laid on tamped cinder or crushed stone, the entire depth of the foundation trench being three feet.

The shed-roof type of construction is used with nine foot studding in front and four and one-half foot studding in back. All studding and rafters are 2 x 4 hemlock or yellow pine. A 2 x 6 girder runs the length of the building supporting the rafters and itself being supported every ten feet by 4 x 4 posts, resting on concrete piers. The plates should be made of 2 x 4 material doubled and joints broken.

All outside walls and roof should be single boarded, preferably of eight or six-inch tongued and grooved yellow pine; white pine can be used, but is much more expensive. The roof and back wall should be covered with a good roofing paper; all joints should be carefully lapped and cemented.

The muslin curtains in the front wall are hinged at the top and can be lifted up. The 3 x 5 glass is hinged at the side and open as indicated on the floor plan. One window in each pen should be so constructed that part of the wall will open when desired, thus making a combination door and window. This will greatly facilitate cleaning and filling hoppers, etc., in an extremely long house.

The dropping boards, perches and nests are best arranged on the back wall, the perches being hinged to the wall so that they may be hooked up when cleaning, the nests being darkened by a hinged door in front which may be let down when it is desired to remove the eggs.

The dividing partition between the units is built of boards and extends from the back wall to within six feet of the front wall; the remaining space is left entirely open. This protects the birds from any drafts when on the roosts. When desired, portable light wire partitions may be used to separate the units.

A large dry mash hopper should be built into this middle partition. If four or five units are built, it is only necessary to have a hopper in the center of each two units, the other dividing partition being utilized for nesting space. This hopper should be constructed with a wooden cover hinging at the center. There is a slatted elevated platform under the muslin front which provides room for the water fountain and grit and shell hoppers.

When the house is completed, a concrete floor should be laid, and should consist of three distinct layers. First, a layer of about six to ten inches of cinders or coarse gravel tamped thoroughly to serve for drainage purposes to keep the soil moisture away from the bottom of the floor. Next, a rough coat of concrete about four inches thick, and over this a finishing coat of two parts of sand to one of cement, troweled smooth and rounded at the corners. Where there is danger of much moisture coming up from below, it is advisable to pour liquid tar between the rough and finished coat of cement, leveling it with a broom before the finish coat is laid.

Such a floor is moisture-proof, rat-proof, vermin-proof, and easily and quickly cleaned.

The following is a list of materials which will be required for building a double unit.

List of Materials Required and Approximate Cost:
Lumber.

Sills .... 6 pieces 4 x 6 by 20 feet hemlock.
Plates .... 8 pieces 2 x 4 by 20 feet hemlock.
Posts .... 2 pieces 4 x 4 by 14 feet hemlock.
  .... 2 pieces 4 x 4 by 18 feet hemlock.
Studding .... 9 pieces 2 x 4 by 14 feet hemlock.
  .... 4 pieces 2 x 4 by 14 feet hemlock.
Rafters ....21 pieces 2 x 4 by 22 feet hemlock.
Frame for nests
  and dropping
  boards .... 5 pieces 2 x 3 by 16 feet hemlock.
Eight inch tongued and grooved yellow pine boards
  for roof, dropping boards, walls and nests.................2,200 square ft.
1 x 2 white pine for curtain frames and trim.................200 linear ft.
1 x 4 white pine for nests.................100 linear ft.
One bundle plaster lath for brood coop.

Nails .... 10 lbs. 20 penny wire.
.... 50 lbs. 10 penny wire.
.... 20 lbs. 8 penny wire.

Approximate cost of the above	$ 75.54
Roofing paper, 1060 sq. ft. or 11 rolls at $3.00	33.00
Four special sash, 3 x 5 feet, at $2.00	8.00
Muslin, 8 sq. yds. at 20 cents per yard	1.60
Hardware, as hinges, locks, tacks, hooks and wire	4.75

Foundation and Floor:

Cement, 35 bags, at 50 cents	$17.50	
Cinders or gravel, 30 yds. at $1.00	30.00	
Sand, 5 yards	7.50	
		55.00
Total cost, not including labor if cement floor is put in the house and cinders and sand have to be purchased		
		$177.89

This gives a cost per square foot of floor space of $0.222.

A cost per running floor of house of $4.44.

A cost per bird, allowing 4 sq. ft. per bird of $0.888.

Adding labor to this at one-fourth the cost of material, the total cost is $222.36, or $1.11 per bird.

Factors Determining Location. Before building the poultry house, great care and considerable time should be given to the study of the best location for the house, and in the case of the amateur desiring to purchase a farm, special care should be given the location of the farm itself. The following suggestions briefly outline the more important considerations with reference to the location.

1.—The location selected for the poultry operations should be within reach of large, and if possible, a variety of markets. This will admit of a more efficient distribution of the products produced.

2.—Rapid and efficient means of communicating with the markets should be available, such for example, as rural free delivery and the rural telephone.

3.—Efficient means of transportation should be present and should include adequate express and freight service within easy reach of the farm. A rural trolley express in the community is a big asset, especially if it passes by the farm gate.

4.—Good roads and slight grades is another point in favor of a location possessing them, for they mean the ability to haul greater loads in faster time and will admit of the efficient use of the automobile for business purposes.

5.—The location selected for the farm should have a rather mild climate and should be free from wide extremes in temperatures. The presence of frequent heavy fogs is undesirable. Locations in low valleys, along river banks or close to the seashore are more apt to be damp and unhealthy, whereas, a higher altitude in a rolling country is always the best.

6.—The soil should be of a porous, sandy nature. There is probably no better soil for poultry farming than a sandy loam, such a soil being well drained and dry, yet being of such a consistency that it holds fertility and will admit of considerable crop production. Such soils are also earlier in the spring, and produce a more balmy, mild climate during the winter.

7.—The laying houses themselves should be placed on sloping land, preferably on the southern slope, with the house facing south or southeast. In this condition they will receive direct rays of the sun the greater part of the day, and thus be much more congenial and healthful.

8.—The houses should be situated in respect to one another with the view toward saving time and labor in caring for the flocks. They should be so planned that when the plant is completed, it will possess a neat and attractive appearance, for this is one of the fundamental principles of modern advertising.

9.—The exact location of the house will be determined largely by local conditions, and these will naturally vary on different farms. Rough, waste land can often be used for the laying stock, the house being located in permanent pasture or wood land which has recently been cut over. If located among forest trees, they should be cut away from round the house to admit of plenty of sunlight entering same.

Size of Flock. The size house to be built will be determined by the size of the flock and the system of poultry keeping which is to be followed. Where egg breeds are kept and egg production is the primary object, the best results from the standpoint of cost of production and products produced can be realized from moderately large flocks where from two hundred to five hundred are kept in single units. Each unit should be given extensive range during the summer, but kept closely confined during the winter. By working on the basis of the large unit, labor is greatly reduced, and the cost of housing per bird is much less than where small, isolated houses, with a capacity of twenty-five to fifty birds, are used. With small flocks, the more detailed and individual attention, which the birds will receive, will doubtless make possible a slightly higher individual production, due to this additional care, but the cost of labor in producing this slight increase is greater than the value of the additional eggs secured. A large investment in buildings of elaborate and costly design is unwarranted. Such buildings rarely furnish as desirable conditions as do houses constructed much more cheaply, but with the health and contentment of the flock, the direct purpose in view.

Environment is one of the great factors in production. The essentials of suitable environment have been outlined. The importance of these factors cannot be too strongly stated. An appreciation of the fact that it is the healthy, contented, well fed, singing hen that lays at a profit will do much towards insuring success from the beginning.

PRACTICE CAREFUL SANITARY PRECAUTIONS AS PREVENTION AGAINST DISEASE.

With chicks in the brooder house, with growing stock on range, and more especially with the laying stock closely confined to the house in winter, the question of sanitation and cleanliness becomes one of considerable moment. Regardless of the type of construction, or the age, or the kind of birds, if they are to be kept free from disease, and in a vigorous laying condition, it is necessary to practice careful and thorough sanitary management. In the case of the laying birds, or the adult stock, this work naturally groups itself into three distinct operations.

The Care of the Droppings. Although one frequently sees dropping pits in use under the perches, yet from the standpoint of sanitation, the dropping board elevated above the floor and directly under the birds, which collects the droppings and keeps them out of the litter and away from the birds, is the best. The dropping boards should be kept covered with some good absorbent. Gypsum or land plaster is excellent, ground loam is good, and ground phosphate rock is frequently used. The use of these absorbents helps to dry out the moisture, keeping the droppings in an odorless and sanitary condition. The droppings should be removed from the boards whenever they give off objectionable odors. It is a good practice to do cleaning at a definite time, probably the most economical practice being to clean dropping boards at least twice a week during the winter. Storing the droppings in an especially constructed shed or in barrels, which are rain proof, is the best practice.

The Proper Care of the Litter. The handling of the litter in the house should have equal consideration with the care of the droppings. The litter, to be efficient, should be coarse, deep, dry, and free from an unnecessary amount of poultry manure. Straw, leaves or coarse hay can be satisfactorily used for litter. There is probably nothing better than rye straw. The litter on the floor of the house should be removed and replaced with fresh, clean material whenever it becomes wet, due to long damp spells of foggy weather, or due to the rain beating in. It should also be replaced when it becomes finely ground and thus loses its ability to hide the grain. Its removal is also necessary whenever it becomes soiled or mixed with large quantities of droppings. If about an inch of dry sand can be kept directly on the floor, and covered with coarse straw to about a depth of six or eight inches, the ideal sanitary conditions will be produced.

Make a General Cleaning. Of equal importance with the previously mentioned sanitary precautions is the necessity of giving the house a complete and thorough cleaning at least twice a year. This means the removal of all the fixtures at least twice a year, such as nests, fountains, and hoppers. It also means the removal of all litter, and the careful sweeping of the inside of the house, including a thorough beating of the muslin curtains. When the house is swept dry clean, it should be sprayed with a good disinfectant solution. The following mixture, known as the New Jersey State Disinfecting Solution has proven very efficient. The mixture is as follows:

5 quarts of lime which has been slaked to the consistency of cream,
1 pint of zenoluem or other equally efficient disinfectant.
1 quart of kerosene.

These three ingredients should be mixed together and completely agitated, after which the mixture can be diluted by the addition of ten quarts of water. The solution should be applied with a force pump with a spray nozzle where possible. When the pump is not available, it can be used as a whitewash. A thorough application of this disinfecting solution will accomplish three different things, more quickly and more easily than can be done in any other way. First, a good coat of whitewash will be applied. If sprayed with force it will reach all cracks and crevices. Second, the disinfectant used will kill all disease germs which may be present in the house, and will also act as a deodorant, making the house smell sweet and clean. Third, the kerosene will help to kill and drive out red mites. After spraying, the fixtures can be returned, fresh litter put on the floor, and the birds returned to a clean house. Body lice and red mites, when present in any number, act as a serious drain on the vitality of the flock. Care should be taken to prevent them from getting established. Kerosene, carbolineum or any good wax perch paint, if applied periodically to the perches, dropping boards, and nests, will prevent the mites from becoming established. A dust box in the house in which is kept a good mixture of fine road dust and sifted coal ashes, in equal parts, will enable the birds to rid themselves of body lice. A clean house with proper protection against vermin will mean more congenial surroundings and healthier birds.

PROPERLY BALANCED RATIONS AND A CORRECT FEEDING PRACTICE ARE IMPORTANT REQUISITES.

A great deal has been spoken and written about the balanced ration, the meaning of the term often being obscure. By balanced ration is meant the daily diet of a bird which contain the food nutrients in the proper proportion which gives her the material necessary to maintain her body, and to produce additional products, such as eggs or meat. The average poultry ration contains five so-called food nutrients, the proportion of which determines the balance of the ration. These food nutrients are protein, carbo-hydrates, fat, ash and water.

The Term Protein is used to designate nitrogenous compounds which are utilized primarily in the formation of lean meat, muscle, feathers, and which goes quite largely into the building up of the albumen in the egg. Protein is supplied to the birds in the form of concentrated vegetable by-products, such as oil meal and gluten meal, and also in certain animal products, such as meat scrap, fish scrap and bone.

The Term Carbo-Hydrates applies to a certain group of food materials which contain carbon, hydrogen and oxygen, some examples of which are sugar and starch. The common source of carbo-hydrates being the grains. Fat is nothing more or less than oil stored in large cells and which is utilized by the birds, in very much the same way as carbo-hydrates are used. The carbo-hydrates and fats in food are utilized to furnish heat to keep the body warm. They also provide heat to furnish energy to enable the bird to move, and in addition to the requirements for heat and energy they are changed over into a compound which is stored up in the body to act as a sort of reserve food supply of body fat.

Ash, as the name applies, is the mineral matter in the ration or the ingredients which will be left after burning. The two most common kinds which interest the feeder are lime and phosphoric acid. Lime is utilized in the formation of bone, and more especially in the manufacture of the shell of the egg. Phosphoric acid is found in considerable quantities in the bone. Phosphoric acid serves a two-fold purpose, first, it builds up bone in the body, and, second, experiments show that phosphoric acid has an important function to perform in increasing the digestibility of other nutrients fed, such as protein.

Water, while not a true food nutrient, nevertheless has many functions to perform. It regulates body temperature, it dissolves food materials and transports them to different parts of the body, and it fills up the body cells, keeping the bird in a plump, healthy condition.

The following discussion of feeding will deal primarily with rations for egg production. At a later place, a discussion of brooding and rearing rations will be given.

Requirements of Rations for Laying Hens. When feeding the laying hen, an abundance of food best suited to maintain vigor, and to stimulate the reproductive organs is necessary if a profitable egg yield is to be secured. There are then two objects in feeding the laying hen, first, to maintain weight and to repair the wastes which normally go on in the body and in addition to this, to supply the nutrients required for the manufacture of the eggs. The following figures give the average composition of the bird's body, compared with the average composition of the whole egg.

	Water.	Ash.	Protein.	Fat.
Comp. of Laying Hen	55.8%	3.8%	21.6%	17%
Comp. of Fresh Egg	65.7%	12.2%	11.4%	8.9%

The above figures point forcibly to the necessity of water in the ration for laying hens. More water, in fact, is required than the bird will normally drink from the fountain, which means that an additional supply in the form of succulent feed must be given. A rather high percentage of protein is noted in the bird's body, and especially in the egg. The dry matter in the egg is one-third protein. This accounts for the necessity of feeding a relatively large amount of beef scrap and concentrated nitrogenous feeds to provide this material where commercial egg production is the object. This table also explains why it is that on the average farm, poultrymen cannot be expected to produce a good yield of eggs on an exclusive corn diet, for corn is not essentially a protein carrier. The large amount of ash in the egg is accounted for by the amount of ashes in the shell. In feeding birds for any purpose, whether flesh growth or egg production, it is a fact that they will utilize the food given, first to maintain the body, and after the body is maintained, an additional supply will be utilized to produce the product, egg or meat, hence the necessity of feeding greater amounts of food as egg production increases in amount.

A great deal of experimental work has been carried on in an effort to determine the amount of nutrients required by the average hen per day when in laying condition. The following table will be interesting to the student who makes a consistent effort to carefully plan scientifically accurate rations.

(Amount of Food Nutrients Required by 100 lbs. Live Weight of Birds per day.)

	Total Dry Matter lbs.	Ash lbs.	Protein lbs.	C. H. lbs.	Fat lbs.	Fuel Value Cal.	Nutritive Ratio.
Hens of 5 to 8 lbs weight..	3.30	.20	.65	2.25	.20	6,240	1:4.2
Hens of 3 to 5 lbs. weight..	5.50	.30	1.00	3.75	.35	10,300	1:4.6

These figures are from the feeding standards derived by W. P. Wheeler of the New York State Experiment Station at Geneva. In compounding a poultry ration, there are a few essential factors which must be appreciated. These are as follows:

A Sufficient Amount of Food Nutrients Must Be Given. Expressed differently, this means that birds cannot lay when poorly fed. It has been found that a bird will not lay without her body contains some surplus fat. It is, therefore, essential that the laying hens should receive an abundance of feed, a little in excess of the immediate requirements. This excess will not be wasted, but stored up in the body for future use.

The Food Nutrients Must Be in the Right Proportion. In addition to having plenty of feed, the proportion of the carbohydrates, protein, fat and other nutrients must be such that it will provide materials in proper proportion for requirements. This ratio or proportion will be expressed by the term Nutritive Ratio, which means the relation of the amount of protein or nitrogenous feeds to the carbo-hydrates and fat. In a laying ration, this ratio is relatively narrow, about one part of protein being given to about four and one-half parts of carbo-hydrates and fat. If a wider ration than this is fed, the bird will put on flesh too rapidly at a sacrifice of egg production. If a much narrower ration is fed, the tendency of the feed is apt to cause birds to break down under the strain, due to its forcing effect.

The Ration Must Contain Much Succulence and Be Palatable. Experiments conducted at many Experiment Stations and experience on successful poultry farms unite in calling attention to the fact that the feeding of succulent feeds, especially during the winter, helps to keep the flock in a normal, healthy condition, whets the appetite and materially increases egg production. A succulent feed is one which contains an abundance of vegetable juices, vegetable juices being a term used to define the juices as they exist in a green growing plant. Common examples of available poultry feeds being lettuce, cabbage, mangel beets, sprouted oats, and any green growing grass or cereal which is in a young, tender condition. The feeding of some succulent material increases the amount of food consumed by increasing the palatability of the ration which means greater efficiency.

The Feeds Must Be Economical and Not Cheap. We have often heard it said that the poultry can eat any kind of feed and it is sometimes the practice to give the birds waste, mouldy feeds which will not be eaten by the other forms of farm live stock. There is no animal which will suffer more quickly from digestive disorders than the fowl, nor is there anything which will upset the digestive system more quickly than sour, mouldy feeds. On the other hand, the cheapest food material is not always most economical, if proper attention is given to the consideration of the quality contained in it. The accurate way of purchasing most feeds, such for example as concentrated protein carriers like meat scrap, consist in determining the amount of digestible food material present, and then knowing the cost per ton, determine the actual cost of such material per pound. Since protein is the most expensive food nutrient to produce, the cost per ton is usually based on the cost of digestible protein. Protein or nitrogen is the nutrient which must always be present in sufficient quantity and the one which most frequently has to be purchased, as it cannot be produced in any great quantity on the farm. The following example shows the method of determining the cost of a pound of protein, and it can be safely said that in purchasing beef scrap, protein should be secured for five cents a pound.

High Grade Meat Scrap containing 50 per cent Protein.
One ton will carry 1,000 lbs. of Protein.
It will cost $50.00 per ton.
Which means 5 cents for each pound of protein present.

Paralleling economy is a question of quality. This is especially significant in the purchase of meat scrap. Two grades of scrap are usually found, a so-called high grade which has been carefully rendered and which contains from fifty to sixty per cent of protein and which can be purchased from $50.00 to $60.00 a ton. There is also a so-called low grade beef scrap which frequently contains from thirty to forty, or even forty-five per cent which costs from $40.00 to $50.00 a ton. It is quite frequently the case that this low grade scrap has been improperly rendered, spoils readily, and may in certain instances contain certain toxic or poisonous materials, which when fed to the birds, results in a high mortality. Special caution should be used in selecting a high grade properly rendered meat scrap. This product is one of the most essential in the egg laying ration, and one of the most costly. The best is none too good.

The Ration Must Be Regularly and Intelligently Fed. Regularity in the care of all kinds of poultry is the primary requisite for the birds are suspectible to a change of routine in management. Any great variations in time or amount of feeding, especially during the winter, when they are working under a high tension, will result in their going off their feed, so to speak, and a reduced egg yield must follow. A definite time should be planned for the feeding of the different rations, and a system so arranged that this feeding can be done each day at the same time, and with the same degree of care. The successful feeder is one who can give to his work careful and intelligent thought and one who is willing to use careful judgment. There are constantly coming up new conditions which from the standpoint of economy and efficiency, require changes in the ration itself, and frequently require considerable change in the amounts fed. No one ration is the best for all conditions and at all times. The prices of food materials are constantly changing, weather conditions, age of birds and other differences which require their corresponding adjustment in the ration are continuous.

Special Supplemental Feeds Necessary. Grit, while not in itself a food material, yet owing to the necessity for a hard substance in the gizzard to grind the grains eaten, becomes a determining factor in the efficiency of the ration. Grit should be kept before the flock continuously in hoppers, a hard, sharp grit being the best. Oyster shell is another very necessary supplemental feed. It is one of the most efficient sources of lime which it is possible to secure, the lime going readily into the formation of the egg shell and providing ash for the frame work of the body. Crushed oyster shell should be kept in hoppers continuously before the laying stock. Charcoal is quite frequently used in the mash, largely for its cleansing and purifying quality. When used, it should be mixed into the dry mash to the extent of from two to not over five per cent. A little salt in the ration for laying hens seems to increase the palatability and doubtless slightly increases the availability of the ration, due to the fact that it has a property of aiding diffusion, which must increase the assimilation of the nutrients. Where the flock is slightly off its appetite, a little salt may bring them back again. When fed, salt should be given in the dry mash and can be used to the extent of about five ounces to one hundred pounds of mash. These factors just described play a very important part in determining the kind of ration to feed and the manner of feeding it.

A Simple, Practical Egg Producing Ration. There are a great many systems or methods of feeding birds. Experience shows, however, that a combination of hopper and hand feeding works out to the greatest advantage. The dry mash should be previously mixed in considerable quantity and fed continuously in large self-feeding hoppers. (See plate No. 5.) Supplemental to this mash, cracked grains and whole grains should be hand fed twice a day in deep litter in the floor of the house. The continuous use of the wet mash system of feeding is not as efficient as the dry mash, for it requires more labor, and is apt to produce, if care is not used in feeding, diarrhœa and serious digestive disorders. Its use where egg production is concerned has almost been abandoned in favor or the safer dry mash method. When, especially in the winter, it is desirable to give birds a tonic, or when it is desirable to get them to eat an additional quantity of the mash, or when for some reason it is desired to force them to a greater production, limited use of the wet mash system must be desirable.

The following is the New Jersey State Dry Mash, and the supplemental rations which are designed for the complete feeding of laying hens throughout the winter. Such modifications as are necessary for summer feeding and for different breeds are also described.

Winter Dry Mash.

Kind of Food.	Amount by weight lbs.	Amount by measure qts.	Cost
Wheat Bran	200	380	$ 3.20
Wheat Middlings	200	240	3.50
Ground Oats	100	100	1.65
Corn Meal	100	95	1.65
Gluten Feed	100	80	1.70
Alfalfa	100	200	1.60
Meat Scrap	200	176	5.50
Total	1000	1271	$18.80

The average cost per 100 lbs. is $1.88.

Prices used are normal and not as affected by war demands.

This mash should be kept before the birds at all times in large self-feeding hoppers. During the moulting season in the fall, it is desirable to substitute oil meal for the gluten meal in the same proportion to hasten the growth of the feathers. As soon as the birds get green grass range, the alfalfa can be gradually omitted and the meat scrap slightly reduced in amount. The extent to which the above mash can be cut during the summer will depend upon the character and amount of the range which is allowed the birds.

The mash is designed especially for the feeding of the Leghorns; when heavier breeds are kept, such as Plymouth Rocks or Wyandottes, especially with yearling or two-year-old hens, the tendency will be for them to taken on access of fat. Under such conditions it is the best policy to restrict the amount of mash eaten by leaving the hopper open during the afternoon only, thus inducing the birds to work during the morning hours for the cracked grain fed in the litter at the morning feeding. The following modification of the above mash will be found very economical for summer feeding where the hens have considerable range and plenty of growing green food.

Summer Dry Mash.

Kind of Food.	Amount by weight lbs.	Amount by measure qts.	Cost
Wheat Bran	200	280	$ 3.20
Wheat Middlings	100	120	1.75
Ground Oats	100	100	1.65
Gluten Feed	50	40	.85
Meat Scrap	25	21	.75
Total	475	561	$8.20

The average cost per 100 lbs is $1.70.

As a supplemental ration to the dry mash, the following grain rations are fed. A scratching ration of whole grain is fed every morning both winter and summer, about nine o'clock, in deep litter. Its primary object, aside from its nutritive value, is to induce the birds to take a considerable amount of exercise. About five pounds of this scratching ration is fed to each one hundred birds on the floor of the house or under some shelter where the litter is dry and where there is protection from cold winds. The scratching ration is made up as follows:

Scratching Ration.

Kind of Food.	Amount by weight lbs.	Amount by measure qts.	Cost
Wheat	100	53	$2.20
Oats	100	98	1.93
Total	200	151	$4.13

Cost of 100 lbs, is $2.06.

At four to five o'clock in the afternoon, depending on season, a night ration is fed, composed of whole and cracked grains, at the rate of ten pounds to one hundred birds.

Night Ration.

Kind of Food.	Amount by weight lbs.	Amount by measure qts.	Cost
Cracked Corn	200	120	$3.30
Wheat	100	53	2.20
Oats	100	98	1.93
Buckwheat	100	66	2.00
Total	500	337	$9.43

The cost of 100 lbs. is $1.80.

It will be noted that by feeding a night ration as outlined, the materials are supplied to keep the birds warm during the night. The above ration is designed for Leghorns. When feeding heavier breeds, it is desirable to eliminate one-half of the cracked corn and to substitute barley for the buckwheat. During the summer months a night ration of equal parts of corn, wheat, oats and barley will supply all the needs for Leghorns. A good rule to follow in feeding the night ration is to give the birds all they will eat in twenty minutes, and then a little more so that there will be some left for them to work on in the morning.

One great advantage of the dry mash method of feeding is the fact that the birds are allowed to balance their indvidual rations in large measure according to their particular tastes and requirements. The feeding of some succulent material in addition to the grain rations is very necessary for the best success.

Providing Winter Succulence. As previously mentioned, the feeding of some succulent material in addition to the above ration cannot be too strongly recommended. Every poultry farmer should grow some mangel beets and thus provide an adequate supply of winter succulence at small cost. (See plate No. 6.) These should be planted in May and sufficient distance left between the rows to admit of horse cultivation. The tops can be harvested and fed in the fall just before the beets are pulled.

The beets themselves can be stored in a root cellar, or in a specially constructed pit out of doors. When no succulent feed is available in the winter, the use of sprouted oats is recommended. The following method of sprouting oats for poultry feeding has been found to be very successful.

The oats should be soaked in water at a temperature of from 60 to 70 degrees F., for about forty-eight hours in pails or galvanized wash tubs, and during this soaking process there should be added from five to ten drops of formalin to kill the spores of mould to insure a clean, sweet feed. After soaking, they are spread out about one inch thick on trays, which are placed in a sprouting rack, seven to each rack, the trays being ten inches apart, and kept at a temperature of from 60 to 80 degrees. (See plate No. 7.)

In from seven to ten days, depending on the temperature, they will have developed sprouts about three to four inches long, as well as a massive root growth, the entire mass being very tender and succulent. The birds will eat this material ravenously. About one square inch of feeding surface is supplied daily to each bird, or what they will clean up quickly. The oats cannot be fed in excess as they are laxative and are apt to produce diarrhœa. The rack shown in Plate No. 7 has a capacity for approximately 500 hens, if kept working constantly, one tray for each day in the week.

Providing Summer Succulence. For the poultryman who is confined to a small area and who cannot give his birds green grass range in the summer it will be found profitable to divide the range or runs into two separate fields, upon which a definite crop rotation can be worked out, allowing the birds to feed first on one yard and then on the other. By planting seasonable crops, such as peas and oats, buckwheat, soy beans and crimson clover, the birds can be provided with a continuous supply of greens from the early spring until the late fall. These crops should be allowed to grow about four to six inches before the birds are turned on them, otherwise they will be consumed immediately and will not last more than a few days. On the other hand, these crops should not be allowed to become tall and woody, in which case they will lack the sufficient property. By rotating the two yards, the birds will be eating up one crop while an additional crop is growing. This practice not only furnishes an abundant material in a cheap form, but the cultivation of the land maintains it in a clean, sanitary condition, which factor is especially important with a lot of birds confined to a restricted area.

The following table shows a simple crop rotation, together with the seeding periods:

Yard No. 1.		Yard No. 2.	
Seeding Time.	Crop.	Seeding Time.	Crop.
April 15th	Peas and Oats	June 1st.	Buckwheat.
July 15th	Soy Beans	Sept. 1st	Buckwheat and Vetch. (Crimson clover in South Jersey.)

A SYSTEMATIC EFFORT TO BREED FOR VIGOR, EGG PRODUCTION AND BODY CONFORMATION.

The aim of every poultry keeper, if he expects to remain in the business, should be to continuously build up his flock by breeding. Where poultry and egg production is the primary object, the breeding efforts will be along two distinct lines. First, to develop the most efficient egg machine which it is possible to breed. This means that his birds must be capable of turning out a maximum number of eggs of good quality during the winter season of high prices. This product must be maintained at the minimum cost for feed and labor. The second object should be to develop a bird for table purposes which will attain a sufficient weight in the shortest possible time, the flesh being of high quality and put on with the least expenditure for feed, thus securing the greatest possible margin of profit.

Paralleling these two aims in breeding should be the continuous effort to breed for vigor and stamina. I would that there were words in the English language which would enable one to express the great importance of vigorous stock. During the time that one is breeding for these so-called ability or commercial characters, he should not lose sight of breed characteristics, and by selection and careful mating, should improve his birds in respect to the body conformation which is required of that respective breed, and in order to maintain uniformity and an attractive appearance about the farm, he should study to fix more permanently a uniform plumage pattern.

The results cannot be accomplishd by promiscuous breeding, but they can be secured by continuous selection and the making of small special matings each year following out a definite scheme of in-breeding and line-breeding. A brief discussion of the method most available follows:

Selection is a Wonderful Opportunity for Improvement. None will deny the fact that variations exist in birds, some good and some bad. The power of selection which the poultryman possesses is a wonderful instrument for improvement. The difference in birds is made possible by variation and by a continual selection of those which possess desirable qualities and propagating these qualities into future individuals. A higher standard of efficiency in the progeny will thus continuously be secured. Careful attention to breeding accomplishes two definite things, it increases the production of individuals, thereby making it possible to secure higher individual records, and secondly, it stimulates the average production of a flock by raising the average of the mass through the elimination of poor producers and the substitution of heavy layers in their place.

Selection should not only be continuously practised in mating the breeding flocks, but it should be the plan to eliminate weak or sick birds throughout the brooding, rearing and adult periods whenever they appear. Fowls which show at any time a lack of inherent ability to resist disease are never a profitable animal on the farm.

Breeding for Vigor. Constitutional vigor, or expressed differently, inherent vitality and stamina, is pictured by the perfect health, the activity and by the vitality which is seen in strong fowls. Birds showing a lack of these are unsatisfactory, both as producers and reproducers. As we expect more and more of the modern hen in the way of production, fowls often break down and the effect is especially shown in future progeny. Much of the low vitality and poor hatching quality in eggs, much of the weakness in brooder chicks, much of the mortality and disease in adult stock, can be traced to lowered vitality in ancestors, due in many cases to the immense requirements for production.

The average hen is expected to lay in a year from four to five times her body weight in eggs. This means one egg approximately every third day of the year. In order to perform this fete of production, she must consume approximately twenty-five to thirty times her body weight in feed. There is, doubtless, no farm animal which is more efficient as a transformer of raw material into the finished product than the hen. The successful breeding for vigor means the appreciation of two sets of factors, first, the lack of vitality, and second, signs in an individual which determine the presence or absence of vigor. Valuable work has been done by a number of our Experiment Stations in studying these factors.

The successful commercial poultryman and the farm poultry keeper who have studied their birds have learned that forcing, due to heavy feeding, or to intensive conditions, if continued year after year, cannot but, in the end, break down the physical strength of birds so treated. They have also observed that inbreeding for a number of generations, without regard for vigor in succeeding generations, intensify the characteristics of low vitality which the original parents possessed. The use of pullets for breeding purposes, due to their immaturity, cannot but result in progeny of small size and possessed of less than their full quota of stamina. Forced feeding during the winter and fall, especially of concentrated protein feeds, has an immediate effect of taxing the digestive system, causing the bird to go off its feed and lowering its energy and physical strength, also the continued crowding of breeding stock into poorly ventilated quarters and the giving to them an insufficient amount of exercise is another direct cause of low vitality. Such conditions will be apparent in the fertility and vitality as possessed by the germ in the hatching egg. Lack of care in hatching and improper range conditions for the growing stock are two other common causes of lack of vigor. Probably the greatest of all causes is the failure of the poultryman to select his breeding stock with great care. Breeding from non-vigorous birds means non-vigorous progeny, whereas the breeding from vigorous birds means vigorous progeny. When mating up the breeding pen, select male birds which show signs of physical strength and superiority, for example, the bird with a bright, prominent eye, with a well-developed blocky body, with an erect carriage, glossy plumage, and bright comb and wattles. The vigorous birds are usually active and spirited in their movements. They range extensively in search of forage. They will be seen to scratch energetically in search of feed. In a great many cases they are the last birds on the perch at night and the first birds off the perch in the morning. In the case of the male, the loudness and frequency of the crow is an indication of physical superiority, while the continual cackle and singing of the female has the same indication. (See plate 8a and b.) It will be evident to any practical poultryman that there is a very definite and fixed relation which exists between the external appearance of fowls and their vitality, hence it should be the aim to systematically select for constitutional vigor at all ages and for all purposes.

Breeding to Increase Egg Production. During the last few years the poultry keepers have been awakened to the realization of the fact that there is a great difference in hens in their ability to lay eggs. We have learned that egg production is not so much a breed characteristic, as it is a character or trait of the particular strain of any breed, in other words, egg production is inherited in individuals, similar to any other body characteristic, and if the poultryman is to continually improve the ability of his females to produce a large number of eggs during the winter when the price is high, he must map out a definite scheme for breeding to increase production. Selection is, of course, the fundamental requisite, however, when breeding for all purposes, but before one can select heavy egg producers, it is necessary to know the record of some of the hens. Investigations have shown us that egg production is inherited primarily through the male parent. For example, the cockerel inherits powers of transmitting high egg production from his mother, if she possessed it. He, in turn, transmits it to his daughters, hence, in order to breed high producing females from high producing females (See plate No. 9) it is necessary to bring this through an intervening generation of male birds. In practice, this means the separation of ten to twenty females, which by trap nest records are known to have produced a goodly number of eggs in the winter. This small number should be given ideal range condition, and during the late winter and early spring should be mated to strong, rugged cockerels. From this special mating, all the cockerels which are to be used in the breeding flock should be produced. A small special breeding pen of this kind will do much to increase the profits from the average poultry farm.

The following facts should be laid down as definite aids in selecting the breeding females. First, it is a recognized fact that if egg production is to be increased, one should breed only from the moderately heavy to heavy producing birds. It has been proven that the most persistent layers are those that lay early in the fall, usually in November, and lay with considerable regularity throughout the winter. All birds, with but few exceptions, will lay fairly good during March and April. The winter production on top of the spring and summer production is what makes the high yearly average. Only matured birds, both male and female, should be used when breeding for egg production. The matured individuals are more prepotent in reference to their characteristics, they lay larger eggs, and thus produce chicks of larger size and possessed of more vigor. Pullets should never be used for breeding. Observations have shown that early producing pullets are very desirable since it insures a good fall egg production. In the early fall it is a desirable characteristic to intensify in future pullets. Cornell University has found by observation that the late moulting hens are usually those capable of the greatest production. It seems, upon careful analysis, that the moult is governed to a great extent by egg production and not egg production by the moult. For example, a bird laying heavily or moderately so in July or August utilizes a great part of the food material which she consumes to manufacture eggs, and thus does not have time to moult, and on the other hand, the hen which moults in July and August has not had to utilize her food for egg production, hence she has diverted it into the form of new feathers. The bird which moults quickly and completely is usually a better producer than the one which extends the moulting period. Birds which consume large quantities of food and exemplify a vigorous appetite should be selected in preference to those which are small feeders. The bird which does not eat freely cannot provide sufficient material for maintenance and production, and hence is bound to be unprofitable even though her cost for keep is less than the heavier eaters. When breeding for egg production, due regard for vigor and characteristics which denote vigor should be considered with those previously enumerated.

Breed For a Uniform Body Conformation. When breeding for so-called commercial or utility characters of egg production, it should be the constant practice to select birds of a uniform body type or conformation. For example, if the poultryman is breeding White Leghorns, select birds and attempt to produce a uniform flock of individuals which have the Leghorn body characters. If Wyandottes, Plymouth Rocks or Rhode Island Reds are the breed, the same suggestion would hold true. The factors of egg production and body conformation, as well as plumage pattern, are inherited entirely independent of each other, and it is possible to produce a high laying strain of Leghorns with any body type which the poultryman might strive to put upon that particular strain, but in view of the fact that for years the American Poultry Association, which is the largest and the official national association, has been striving to definitely fix and breed characters, we will be going backwards rather than forwards if this constant effort to produce birds which have the characteristic breed type is not closely followed. Such a uniform flock is neater in appearance, produces a more uniform product and are by far the most profitable investment. In conclusion, it might be said that breeding is, in itself, largely the practicing of systematic selection. The following motto is only too true and should be set down in a prominent place in the poultryman's practices.

GOOD HENS MEAN BETTER HENS.

USE THE PROPER METHODS IN HATCHING AND REARING.

The time of year for hatching chicks which are to be reared as future pullets, or which are to be sold for broilers, should be carefully considered. A few weeks too early or a few weeks too late may mean the difference between a profitable winter production in the case of pullets, or in the case of broilers it may mean the difference between from thirty cents a pound to fifty cents a pound on the market. Late hatched pullets grow slowly during the summer, owing to the fact that they do not get a good start, and hence do not come into maturity in the proper time in the fall and do not get under way in egg production before winter shuts down. On the other hand, if they are hatched too early, as for example early February and March, they will doubtless moult in the late fall which will cause a lower production during the balance of the winter. The exact time for hatching will be determined largely by the type of bird kept. The American general purpose breeds are characterized by slow growth, and hence must be hatched earlier than the light active Mediterranean breeds which mature in from one month to six weeks shorter time. The Leghorn and birds of their type are best hatched about the middle of April. Where it is necessary to bring off more than one hatch, they can be safely brought off from April 1st to May 15th. This will give them between five and six months in which to mature, and thus be in laying condition by October. Heavier breeds will usually do better if hatched from the middle of March, and not later than the last of April.

Another factor which affects the time of hatching is the condition of the range upon which the chicks are reared and the method of feeding. Youngsters which are provided during the summer with an abundance of range providing shade and green food, and plenty of nutritious food material, will make a more rapid and uniform growth than flocks which are crowded into small bare yards during the same time.

The hatching egg should be carefully selected in order that they may be uniform in shape, size, and as far as possible in color, they should also be strictly fresh and of normal shell. If these characteristics are chosen the tendency is for the progeny to lay a more uniform product.

Experience shows that eggs decrease in their hatching power the longer they are held, and it is never safe to hold hatching eggs over three weeks. When it is necessary to hold them for even a short time they should be placed in a moderately cool temperature, between forty and fifty degrees being the most desirable. They should also be turned occasionally to keep the air cell from becoming misplaced. If possible, it is well to stand them on the end, leaving the air cell uppermost. The character and quality of the chicks resulting from the hatch will depend in large measure upon the condition of the hatching eggs placed in the machine. Too much care cannot be expended in keeping the eggs in a normal condition.

Artificial Hatching. The type of incubator which it is best to use will be governed largely by the number of birds desired. The poultryman keeping a few fowls for family use, not hatching more than one hundred chicks each year, will undoubtedly find it more profitabe and more interesting to use the natural method, or the sitting hen. Again, where a large number of chicks are desired, say from two to three hundred and often including many thousands, artificial incubation has been found to be more practical. Artificial incubation, if properly carried on, will produce just as vigorous chicks as can be secured by hatching with hens. In addition to this, the work can be accomplished with less labor and at the proper time of year. The artificial process is entirely under the control of the operator, while with the natural method one must be governed largely by the wishes and likes of the sitting hen. When determining on the type of machine to purchase, a few extra dollars spent on a reliable well built standard machine is well warranted. The directions which come with the machine, regarding its operation, should be followed carefully. The mammoth incubators, which are a product of recent years, have proven to be very efficient, especially desirable where custom hatching can be done in addition to the requirements on the one particular poultry farm. The incubators should always be located below ground, or in a building which is at least partially below ground. Such a location offers protection from changes in weather conditions, insures an abundance of fresh air, and makes possible the supplying of an adequate moisture content.

Operating the Incubator. Success with any incubator, whether lamp heated or coal heated, requires, first of all, a working knowledge of the principles of incubation, and experience in running that particular type of incubator. With the lamp-heated machine this includes the proper care of the lamp, its regular daily filling, the trimming of the wick, and the maintaining of the flame at a sufficient height to furnish the required amount of heat without smoking. With coal-heated mammoth machines this means great care in tending the fire to see that a uniform bed of coals is maintained, and that the temperature devices are so regularly adjusted as to maintain a uniform hatching temperature of about 102 degrees on the egg. The temperature should not vary either way from 102 degrees, and special attention will have to be given the regulating devices on or about the fifteenth day, as the temperature will have a tendency to rise, due to the formation of animal heat from the germ within the egg. The eggs must be turned twice a day from the third day until they first show signs of pipping. They should be cooled regularly, the length of time will depend in considerable measure upon the period of incubation and upon the temperature within the incubator cellar. The cooling process can be continued for longer intervals as the hatch progresses. In a moderately warm cellar (60 degrees), in the late spring, cooling as long as twenty minutes during the last week of the hatch and usually not longer than the time required for turning during the first week of the hatch is a safe rule. A good test is to cool the eggs down so that they will still feel slightly warm to the cheek or back of the hand. Careful observations upon hatching have shown a great need of moisture. The final results of the hatch, as measured by the vigor and number of chicks, is determined to a great extent by the moisture conditions in the machine. The amount of moisture required will vary with the season and the condition of the room in which the incubator is located. Experiments at our own station at New Brunswick have shown that during the latter part of the hatch, a relative humidity in the machine of from fifty-five to sixty-five degrees is best. These same experiments show that this high moisture content, which must be artificially maintained, resulted in a larger percentage of hatch, in fluffier chicks, in a more uniform hatch, in less dead in the shell, and showed in every respect the need and value of moisture.

Moisture should be supplied to the average incubator in abundance by increasing the moisture content in the air of the incubator room and by sprinkling the eggs with warm water three or four times during the last week of the hatch. Plenty of moisture and proper hatching conditions, if given eggs containing strong germs, will result in good hatches and few dead chicks in the shell. The question of sanitation in the incubator is a very important one. There is the danger of disease germs which might be carried into the machine by the hatching chicks, also the danger from filth being carried over from hatch to hatch. This can be entirely eradicated by washing and spraying the machine between each hatch with a five per cent solution of zenolum and allowing the machine to air from eighteen to twenty hours before filling it for the next hatch. (See plate No. 10.) This sanitary handling of the incubator between each hatch will at least prevent the spread of infection and serve as insurance against the transmission of white diarrhœa from one flock to another.

Artificial Brooding. A possible serious loss to the poultryman is death in the brooder. This loss can in large measure be avoided by selecting a suitable brooding system which will maintain proper environmental condition, and in addition to this, by providing the chicks with a suitable feed ration. For the small poultryman, brooding only two or three hundred chicks, the small outdoor brooder of fifty or sixty chicks capacity may be satisfactory, but for the commercial poultry farm, or for the farm flock where three hundred chicks and upwards are reared. the so-called colony brooder stove will be found very desirable. (See plate No. 11.) These are a recent product, having been put on the market within the last two years. A great many different types are available. One which is substantially constructed and has a rather large metal reflector which can be raised and lowered, and which is provided with an accurate, simple method of regulating temperature should be satisfactory. From three hundred to five hundred birds, (never over five hundred), can be placed under one of these hovers, and a brood of from eighty-five to ninety-five per cent should be secured. The commercial plant doing considerable winter brooding, especially where winter broilers are produced, will find that an intensive brooder house, with a central heating plant and pipe running under the hovers, will have certain advantages. The colony brooders seem to be, at the present time, the most economical solution of the brooding problem. They should be placed in a relatively large house, never smaller than 12x14 feet, preferably of two rooms, one room containing the heater or hover in which a fairly warm temperature is maintained, and an adjacent room in which the chicks can be fed and take their exercise. This colony system of brooding admits of giving the chicks from the beginning considerable range, it cuts down the cost in permanent equipment and labor, it also enables the use of the house continuously throughout the year, for after the chicks have gotten old enough to do without heat, the stove can be removed and the house used for a colony house during the growing period, in which the pullets can be left until they attain maturity. During the winter the houses can be used for a short period for special breeding flocks.

Care of the Brooder Chick. The chicks should be left in the incubator from twenty-four to thirty-six hours after hatching, thus allowing them to dry off and become strong on their feet. (See plate No. 12.) When moved to the brooder, care should be taken that they are not chilled. Before the chicks are placed in the brooder, it should be thoroughly cleaned and disinfected. Clean fine sand should be put over the floor to a depth of one-half inch, and over this, clean cut, alfalfa or clover should be placed. The lamp should be started and the hover heated up to a temperature of about ninety-eight degrees. A high hover temperature is not desirable, as it lowers the vitality of the chicks and makes them tender and more easily injured. After the hover has been tested for about twenty-four hours it is ready for the chicks. When moved to the brooder, any under size, deformed or crippled should be immediately killed. Such chicks are never worth the food required to maintain them. From this time on through the growing period, a continuous selection should be made, weeding out any chicks which show a marked tendency toward lack of stamina, for such birds never make profitable market poultry, nor as females, never produce a profitable dozen eggs.

The hover temperature should be maintained at about ninety-eight degrees during the first few days, after which it should gradually be lowered five degrees a week until at from four to six weeks of age, the chicks can be weaned entirely from the heat. The chicks are the best judge of the brooder temperature and the heat should be such that it will not cause them to chill and crowd, neither will it be hot enough to cause panting. The brooder should protect the chicks from drafts and provide against crowding. For these reasons, the circular hover with a central heat drum seems to be superior.

Feeding the Brooder Chick. The proper feeding of the artificially brooded chick is important, since the digestive system during the first four weeks is very delicate and easily upset by an improper feed. The general practice should be to get the chick through the first few weeks of its growth without forcing, allowing it to develop a vigorous constitution with a good body growth, and after that time it can better stand forcing for a rapid meat growth when desired.

When planning the rations for the youngsters and when determining on the method of feeding, it is important to appreciate that the first feeds should be easily seen and should contain much nutriment. It is also well to practice a restricted or retarded early feeding in order that their delicate digestive organs may not be overcrowded. Grit and shell is an important essential in the chick ration and fresh water should be provided in a large amount. Dry cracked grains are safer, for the first few weeks at least, than wet mashes. Wheat bran is an important addition to the feeding practice, as it contains ash, is slightly laxative, and is relished by the birds. Ash in the form of phosphoric acid can be secured in the form of dry ground bone, and is an additional essential. During the early part of the feeding period the chicks should be fed little and often, and should be kept busy and hungry between feedings. Some animal protein in the form of meat scrap is necessary. Sour milk in a loppered condition is very desirable, as it not only furnishes much food material, but the lactic acid present acts as an internal disinfectant. A continuous effort should be made to practice clean feeding, for nothing will upset the digestive system quicker than sour and mouldy feed. The following feeding practice embodies the above principles, and if carefully followed, will result in good sturdy chicks:

First Eighteen Hours in Brooder. Grit, shell and water, with short-cut alfalfa and sand on the brooder floor.

The Day Following. Pinhead oats or oatmeal, three feedings.

The Next Five Days. Feed the following cracked grain ration, on the brooder floor, five times daily, feeding what they will clean up between feedings.

> 40 lbs. Cracked Corn.
> 40 lbs. Cracked Wheat.
> 20 lbs Pinhead Oats.

Supplemental to this ration, hard-boiled eggs once daily, sprouted oat tops, twice daily (small amounts), may be fed.

Seventh Day. Start feeding wheat bran in small hoppers, leaving it before the chicks two hours, and omit the noon grain feeding.

Eighth to Fourteenth Day. Bran all the time in hoppers, and cracked grains four times daily.

Second to Eighth Week. Keep the following dry mash always before them and feed cracked grain three times—morning, noon and night.

Chick Dry Mash.

Kinds of Food.	Amount.
Wheat Bran	50 lbs.
Gluten Feed	10 lbs.
Corn Meal	10 lbs.
Ground Oats	10 lbs.
Beef Scrap	10 lbs.
Granulated Bone	10 lbs.
Total	100 lbs.

Hardening Off Process. In order to get the chicks in condition for removal to the range, it is necessary after the second week to practice a hardening off process. This should be gradual and consist of lowering the temperature with the idea of doing away with an artificial heat entirely in from three to six weeks, depending upon outside weather conditions. The best practice is to reduce the artificial heat until it can be entirely given up, then gradually to raise the hover until it can be removed and replaced with muslin-covered frames, having them hung to the hover wall, gradually raising them in front a little each night until the chicks become used to their absence. It is impracticable to take them from a warm heated brooder house and put them into a colony house unless they have been accustomed gradually to the change. The idea should be to get them on the range as soon as possible. When they are four weeks of age, the sooner they can be gotten out into the cool temperature in large, well ventilated quarters, with free range, and abundance of green food and access to the ground, the better they will grow, and the hardier and more vigorous will they be at maturity.

This hardening off process is especially desirable with Leghorns, as their close feathering makes them susceptible to cold weather, and when not properly weaned they pile on top of one another to keep warm. This usually results in the death of many and a loss of vitality to the others. The most of the losses in brooding young chicks are due to the following causes:

1. Crowding and subsequent death, caused directly by a too low brooder temperature.

2. A derangement of the digestive system, resulting in diarrhoea, and usually caused by wet, sloppy, early feeds.

3. A loss of vitality and stamina, due to overheating.

4. The tendency which chicks, especially Leghorns, have of devouring one another. It is commonly called cannibalism, develops chiefly in large flocks, and is due to an insufficient amount of animal material in the ration. The remedy should be to remove all birds which have been attacked or any which show signs of blood. A dry mash composed of equal parts of meat scrap, dry ground bone, oyster shell, and wheat bran, should be provided in an open dish. This will correct the ration, and with care the habit can be stopped.

5. A contagious disease known as white diarrhoea. Where the chicks are carried off during their early growth in large numbers, it is often caused by an infection which may be inherited by the young chicks, the disease in its chronic form being found in the ovaries of their mother. The best way in such a case is to completely disinfect the brooding quarters and pro-

vide the chicks with an abundance of sour milk to drink. The germs of white diarrhœa are easily killed in a dilute acid. The only way to avoid future epidemics is to trapnest all hens and find out which are infected. They should be killed when detected.

Care of the Birds on Range. After the chicks are weaned and placed on the range, the aim should be to induce a continuous growth throughout the summer. Any checks or setbacks which they might be subjected to, due to improper feeding or care, will result in irregular maturity and lack of uniformity. There are two factors aside from their inherited characteristics which affect proper maturity. These are environmental conditions and food supply.

Environment plays an important part, as the best bred chicks, possessing all other desirable characteristics, if not given ideal conditions in which to grow, will not be allowed to exercise or develop inherited traits to the fullest extent. These conditions are as follows:

1. One should not attempt to grow young stock on restricted range, for they will not make a satisfactory growth, due to a limited supply of green food and lack of exercise. Free range conditions should be provided rather than intensive methods should be followed.

2. Shade should be provided in abundance. Trees are ideal for this purpose, orchards being especially desirable. If trees are not available, corn or sun-flowers can be planted, and in the absence of either of these, artificial shelters of burlap over wooden frames should be provided.

3. An abundance of green succulent food material is very necessary. If free range conditions are provided, this will be found in abundance. Where it is necessary to grow a large number of chicks on a limited area, **the plot of land should be** divided into two parts, colony houses being placed approximately through the center of the field and a portable fence placed on one side of the houses and later moved to the other side. This will permit rotating the two areas and growing desirable crops to supply succulence. Peas and oats can be sown early in the spring, followed by rape and later by buckwheat. In the

fall, wheat, rye, vetch and clover, any, or all, can be sown to provide a winter crop and furnish early greens the following spring. (See plate No. 13.)

4. The type and size of the house in which the chicks are placed bears a close relation to their growth. Fresh air is the limiting factor. The chicks are only in the house or shelter during the night, and all they need is protection from wet weather and enemies. An ideal summer growing house is one about 6 x 8 feet on the ground with a shed-roof, the front being six feet and the back four feet high. The door can be placed in the center of the front with a long, narrow muslin opening on either side of the door. The lower half of each side wall should be made in the form of a panel, hinging at the top to allow it to be opened out and up. When these two sides are opened, a free circulation of air through the house keeps the birds cool and comfortable on warm nights.

The chicks should be given an abundance of house room, since crowding stunts their growth and results in many weakenings. The colony house 6 x 8 feet, as described, will house from 75 to 100 chicks at 5 weeks of age, and after the male birds are separated at about 10 weeks, it will easily accommodate the 40 or 50 remaining pullets through the balance of the season. In managing the growing stock, personal attention should always be given to their wants, especially when young. Considerable attention is required to see that all are gotten in quickly from sudden showers, that they find their proper quarters at night, and are protected from rats and other enemies. We must remember that these growing pullets are the machines which next year are to consume the raw product, food, and in return give us the finished product, eggs. If the machine represents a high degree of perfection, we can expect it to utilize the food material to better advantage. Perfection in development can only be attained by providing free range and fresh air.

The method of feeding the growing stock is not complicated. The practice should consist of having the food before them all the time so that they can balance their own ration. They will usually take sufficient exercise if given plenty of range, hence the common practice is to feed a well balanced dry mash in large self-feeding hoppers, and supplement this mash with a good, cracked grain ration. The same dry mash should be used that is recommended after the second week in the brooder. In addition to this, a ration consisting of equal parts of cracked corn and whole wheat should be fed twice a day about the range.

This method of feeding will allow the chicks to balance their own rations and will give the weaker ones a constant supply to which they can have access when they are crowded away at grain feeding time by the larger ones. Dry mash in self-feeding hoppers will tend to equalize growth and produce a more uniform flock at maturity, while the feeding of cracked grains entirely will tend to exaggerate and constantly increase any difference in size which may exist. Large self-feeding hoppers, holding from 200 to 300 pounds of mash, can be constructed and placed at frequent intervals around the range. This will facilitate the feeding and make the mash available to all the chicks.

Care at Maturity. Where possible, it is a good practice to place the pullets in their laying quarters at least one month before the flock is expected to gain maturity, for two reasons:

First. Birds are especially susceptible to changes of environment. By giving them time to get acquainted with their future home, retardation in growth is avoided.

Second. If for any reason a maturity is delayed, it is quite possible by having the birds closely under observation, to hasten or retard their ultimate maturity by the feeding of forcing or retarding mashes. In this way it is possible to bring late hatched birds to maturity from three to five weeks earlier.

It is best to have the pullets mature before cold weather begins in the fall. This will be in October for North Jersey and November for South Jersey. This means that the flock average should be established by December 1st, if a profitable yield is to be obtained during the rest of the winter. In other words, it will not be possible to greatly increase the egg yield after that time. An eight per cent. egg yield on or about December 1st will mean a low production during the winter, but a thirty per cent. yield at that time can be kept, with little trouble, well up or above that point, until the following spring, especially will this be true if proper housing and feeding conditions are furnished.

INSURE A HIGH QUALITY IN MARKET PRODUCTS.

It has long been an appreciated fact that prices paid for poultry products vary largely according to quality. There is always a demand at remunerative prices for high grade eggs and dressed poultry. If the poultry keeper is to be sure of finding a ready market, his aim should be to produce products of the best quality it is possible to secure. This means in the case of eggs that they should be produced by healthy birds kept in suitable environment, the nests should be roomy and should be provided with plenty of clean litter. The eggs should be collected frequently, at least once a day, and broody hens should be shut up so that they cannot remain on the nests and cause a development of the germs. On wet, muddy days, the birds should be confined to the house so that they cannot soil the eggs and the nest material with muddy feet. After collection, the eggs should be placed in a clean dry room where they should immediately be graded and shipped. The average wholesale market recognizes two distinct grades of eggs, medium to large, and small. During the fall of the year, or when the pullets start laying, all small eggs should be put in a case by themselves and shipped as such. This practice will result in a higher price being secured for the normal eggs, whereas if the small ones were included with them, it would result in a lower price for the whole shipment. Where eggs of two colors, brown and white, are produced, they should be shipped separately, for the New York market pays a premium of from three to eight cents for the whites over the browns. (See plate No. 14.) When grading the eggs, uniformity should be the main thing sought for. Uniformity as to size, shape and color are paramount. No dirty eggs should ever be shipped to market, for they will lower the selling price of the whole shipment and the poultryman's standing will suffer. Special care should be used in packing. Regulation egg cases of thirty dozen capacity are the most economical for wholesale trade. If used for the second time, the cases should be renailed before filling, and all broken or flimsy containers should be replaced by new ones. A layer of straw or excelsior should be placed in the bottom of the case to break the jar by permitting a certain springiness to the contents. The egg should be placed in the fillers, small end down, no extremely large eggs being packed, as they are apt to become broken and soil the other eggs in the shipment. When all the eggs are in, a thin layer of straw or excelsior should be spread over the top before the cover is nailed on. It should be nailed securely at each end, but not in the middle. This leaves room for spring, and prevents much breakage in transit. When shipping to a limited or a private trade, pasteboard containers that hold from one to six dozen work very satisfactorily. Stencilling the shipping cases or cartons with the name of the farm and the brand has proven to be a very remunerative practice, as it enables one to build up a trade for his particular product. This trade will be secured, if each time care is given to see that the quality is maintained. Eggs should be shipped by express as soon after they are collected as is possible to make up a complete package. The holding of eggs for higher prices, especially in warm weather, if only for a week or two, will cause their quality to deteriorate. The air cells will become enlarged, due to evaporation and a small price will be received for them. Experiments conducted at Purdue University show that 17 per cent. of the eggs shipped to wholesale markets have no commercial value after arrival, due to their being exceptionally dirty or broken, due to the development of the embryo, or due in some cases to their being held so long that they have become rotten or mouldy. The following advice is given in regard to the prevention of unnecessary losses in market eggs. Eggs for market should weigh from one and one-half to two pounds per dozen. They should be uniform in size and should be free from dirt or stain. They should not be washed, but should be dry cleaned if possible. Market eggs should have strong round shells and should be strictly fresh, being not over five days old. They should be laid in clean nests, and should be gathered often. Market eggs should never be taken from an incubator or from stolen nests. From the time they are collected until shipment, they should be kept in a clean, dry place.

Infertile Eggs Are Valuable for Market Purposes. During the early spring and hot summer, much loss in market eggs is due to germ development which is only possible in eggs which have been produced by the pens containing male birds. This germ development is brought about in large measure by keeping the eggs in too warm a place, such as near a fire or exposed to the direct rays of the sun while en route to the market. A broody hen sitting on the eggs or irregular gathering of the eggs is sometimes responsible. The production of infertile eggs for market will tend to eliminate all danger from this source, therefore, its advantages over fertile eggs are many. In the first place, they do not hatch, for they contain no germ which is capable of development. They withstand heat and thus bear shipment well. They are easily preserved since they are slow to decay. They are the best for cold storage or for home preservation. They are less costly than infertile eggs from the standpoint of production, because no male birds are required. They are also produced just as abundantly as are fertile eggs. There is probably no one thing which the poultryman can do which would so improve the quality of his market eggs as the production of infertile eggs.

Dressed and Live Poultry. Both live and dressed poultry, when shipped to market, should be graded and selected as carefully as should eggs or any other poultry products. They should be of designated weight and as uniform as it is possible to have them. (See plate No. 15.) The eastern poultryman, in the shipping of live poultry, must compete with western stock which reaches here in carload shipments. He can compete with these shipments very favorably if his birds are in a ripe, plump, good condition of flesh. When either live or dressed poultry are to be sold for table purposes, quality is the primary consideration, and quality can only be secured through proper feeding. When feeding for flesh or for finish, as is the case with broilers or roasters, or capons, it is usually necessary to slightly restrict the exercise taken by the birds. The rations should be rich in carbohydrates and fat, much of the feed should be fed in a moist condition and plenty of grit and meat scrap should be provided. The most important essential in feeding for finish and quality is to keep the birds growing as rapidly as possible, yet maintaining a good appetite. In the majority of cases, these conditions can best be brought about by fattening the birds in relatively large flocks in confined yards. Crate fattening may be desirable with roasters, especially where large numbers are fed. In the hands of an amateur, however, the crate fattening results in no material gain, whereas flock or pen fattening usually results profitably.

Finishing Broilers. When finishing broilers for market, the object is to make them a little plumper and heavier in a short time. The following ration is recommended for this purpose:

Broiler Finishing Mash.

Kind of Feed.	Amount.
Corn Meal	25 lbs.
Ground Oats	25 lbs.
Beef Scrap	15 lbs.
Granulated Bone	10 lbs.
Wheat Middlings	25 lbs.
Total	100 lbs.

The above mash should be mixed with skimmed milk and fed in a crumbly, not sloppy condition. It should be fed in troughs and kept sweet and clean. In feeding this mash once a day, as much should be fed as the birds will clean up in twenty minutes to one-half hour. It should be supplemented with a grain ration of equal parts of cracked corn and wheat, which should be fed morning and night. When finishing roasters or capons, it is well to confine them so that the exercise will be materially restricted. The following ration will produce excellent results, either fed to the flock as a whole or used in crates. This mash mixture should be moistened with skimmed milk. If none is at hand, water will answer. It should be fed in pans or water-tight troughs. It should be fed two or three times a day, only enough being given so that the birds will clean it up and be hungry for the next feeding. When crates are used, it is well to keep them darkened during the non-feeding periods, opening them up during the feeding time. This will induce less activity and thus conserve more food material for flesh formation. The following is the fattening mash:

Kind of Feed.	Amount.
Corn Meal	40 lbs.
Wheat Middlings	20 lbs.
Ground Oats	20 lbs.
Beef Scrap	20 lbs.
Total	100 lbs.

The sooner the poultryman appreciates that quality in poultry and eggs means higher prices and a bigger demand, and as soon as the New Jersey poultryman realizes that this is the only way which he can successfully compete with western grown meat, the more attention will be given these factors.

THE DEVELOPMENT OF THE BEST METHOD OF MARKETING.

For a number of years, organized poultry interests in the State of New Jersey have been making consistent efforts toward co-operating along buying and selling lines. As a result of this effort a scheme has been devised which is now in operation and working very satisfactorily, whereby members of the New Jersey State Poultry Association can co-operatively market their eggs. It is suggested that all poultrymen in the State who ship eggs in case lots and who feel that they should be able to improve marketing conditions, will find much benefit to them in this scheme. The following extracts from previous publications dealing with this scheme will call attention to the scope, plan and outline of the project. (See plate No. 16.)

Purposes and Scope. We should not be led astray into thinking that although our plan may ultimately include the marketing of all New Jersey produced eggs that now go through wholesalers' hands in large cities, this plan will have any effect upon the market price of the general supply. We should remember that although our State is known as the poultry State, we supply New York City with only one or two per cent. of its total receipts. On the other hand, our eggs are in a small class, and if only first-class, fresh-laid eggs are considered, New Jersey is a large factor. Especially is this true of white-shelled eggs. The purpose of this plan and the need of the average poultry raiser in the State, as the committee understands conditions, is a more efficient method of supplying particular people in large cities with the finest quality of eggs which can be produced. This includes the elimination of loss to a large degree, from improper packing methods, little or no grading, and wholly unsatisfactory systems in distribution. Surely the New Jersey poultryman does not want to compete with the western producer in the production and selling of only fair quality eggs. Because of our nearness to the greatest market section in the country, we have certain advantages which remain for us to profit by if we will. Are we going to neglect our birthright by continuing our slipshod methods of marketing? The demand in New York City, which we should remember is the greatest egg market in the world, is for food products of superior quality and appearance. We can furnish them the best grade of eggs at less cost and greater profit to ourselves than can poultry raisers of any other district. We can, but in many cases we do not. The average producer is as ignorant of market conditions as is the average market man in a large city of any line of farming. We are producing quantities of eggs of the quality in such demand, but they never realize what they should for us because of our inefficient methods of marketing. The persons who demand this high class of products not only ask for freshness, quality and appearance, but they desire to feel absolutely sure that they can depend always upon the uniform high grade and that there will never by any chance be a "mistake." They want to feel sure that the supply is of sufficient volume to furnish them at all times. If any one of these factors is lacking, this class of trade is diverted. The committee believes that only through co-operative efforts of many careful producers can this trade be permanently secured. The possibilities of this plan are almost unlimited. The task at hand is to get it established. It is one step, and we believe an important one, in the work, which the many poultry organizations, with the aid of the State and Federal Departments, are striving to accomplish, namely, to develop the poultry industry into a profitable legitimate business, capable of returning fair profits to those interested in it and who are willing to give it close study and well directed management.

Outline of Plan. The plan includes the selection of reliable, experienced dealers in large cities to act as our agents. We are starting with one in New York City. Instead of starting under a handicap by setting up a new store with inexperienced operators, as has often been tried, we recognize that the selling end of the business is intricate, that there are many conditions which exert a vital influence upon the success or failure of such an enterprise that only experienced men can successfully deal with.

All members of the local association who are members of the New Jersey State Poultry Association are privileged to send their eggs to this agent. This insures uniformity in the product as it reaches the trade, and does away with all detail work which some producers have been forced to take up.

The agent will carry on the business of selling in his own way, but under the supervision and with the full co-operation of the State Association.

Returns will be made on the day of sale and all possible energy will be used to move the eggs rapidly.

Arrangements can be made by individual shippers to have a part or all of their eggs stored during seasons of heavy production at consequent low prices. The storing facilities will be of the best, and our agent is well acquainted with this end of the business.

Requirements to Ship. The only requirements made of those who desire to take advantage or who wish to support this project are: that first they must be members in good standing of the New Jersey State Poultry Association; and secondly, that they will so manage and care for their flock that only eggs of high quality and flavor will be produced. The services of the Poultry Department of the State Experiment Station are available for suggestions and methods of practice.

The secretary of any local association can secure through the committee "Shippers' numbers." Shipping tags will be forwarded to the shippers to use on each shipment. This tag marker with a shippers' number insures its recognition by the agent and the shipper will be credited accordingly.

Small eggs, or those of poor shape, etc., will be disposed of by the agent through suitable channels, but shippers should realize that this association is endeavoring to establish a reputation for excellence and that high grade goods are going to receive special attention.

Upon complaint to the committee by the agent, a shipper's number can be cancelled if, upon investigation, such action seems to the best interest of the association. It sometimes happens that eggs from a flock will be of poor flavor on account of unwise feeding or some other cause. This cancellation clause is inserted to protect the trade.

Duties of the Agent. The agent will receive all eggs from shippers bearing an authorized number. It is urged that where eggs are shipped by express, that they be shipped collect. The agent will then pay the charges, deduct the same when returns are made. This sometimes facilitates the collection of damages, charges, etc.

Each shipment will be graded and candled separately and the shipper given credit for the following grades: first whites and first browns will be large, clean eggs of the highest quality. They must weigh twenty-four ounces to the dozen. The whites must have chalk-white shells and any tinting or color will place them in the class of browns. In warm weather, eggs to come in these classes must not be over a few days old. It will be found that conditions under which eggs are kept will have more influence in determining the quality than mere number of days. Seconds will include all small, shrunken and dirty eggs which may be received. Many times large numbers of small eggs are produced on the best farms from flocks of pullets. Good markets can be found for these in the large cities. Shrunken eggs are those in which the moisture has evaporated more or less. This condition is often found in eggs held for several days or more. It should be guarded against. When a small shipper is receiving only a few eggs a day it may become impossible to fill a case or even half a case often enough to prevent this condition. This matter should be discussed by the local association with the idea that two or more raisers combine and ship together. In this way frequent shipments can be made, which is the only way in warm weather to insure the best results. Checks is a market term applied to all eggs cracked or broken.

Complete accounts must be kept by the agent, which will be open to the association's committee at all times. These accounts must include each sale, the purchaser's name, price paid, etc.

The agent will endeavor to develop a large trade for our best eggs packed in one-dozen cartons. Our New York agent is to put a special man on this work.

All eggs not sold in cartons will be sold in standard thirty-dozen cases.

Summary.

A few of the interesting features of this co-operative project may be briefly summarized as follows:

1. No additional work or inconvenience necessary on the part of the producer.

2. The advantages to the producer arising from having a special interest in the disposal of the eggs and a control over methods pursued. This should lead to quick returns, a perfect understanding between producer and distributing agent, etc.

3. The highest price possible to secure where such limited expense in distribution is incurred.

4. The possibilities of building up better marketing facilities year by year by reason of experience gained and value of reputation secured.

5. A feeling of confidence and security in knowing that our eggs are being sold to the very best advantage, and that we know just how they are sold and are receiving every cent that belongs to us.

6. The advertisement, publicity and prestige which will be gained for the poultry business in New Jersey.

Note.—Anyone interested in becoming better acquainted with this scheme or in securing a shipper's number should write to A. L. Clark, Agricultural Experiment Station, New Brunswick, N. J., who has charge of the details of this co-operative egg marketing project.

APPLY BUSINESS PRINCIPLES TO THE MANAGEMENT OF THE FLOCK.

Agriculture, and especially poultry keeping, is just as much a business as any manufacturing, mercantile or trade occupation.

If it is to succeed and to be put on the same substantial basis as other business enterprises, it should be run on true business principles. Expressed in a simple phrase, this means that the poultryman must know all about the details of his business, both financially and otherwise. He must know, in other words, at all times, the details of the organization, production and distribution.

These results can only be obtained when simple, but complete records are kept, the information of which will enable him to determine at a glance the actual cost of operations, together with the incoming and outgoing money. Financial success in managing the poultry flock depends not only on the abundance of production, but on the cost of this production. Once the product is produced, the profits of the business depend largely upon the ability of the poultryman to dispose of his products at the greatest advance over the cost of production.

Keep Simple Egg Records If Nothing More. The use of especially printed or ruled sheets in the laying pens upon which can be recorded the approximate amount of food fed, the number of eggs laid, the number of hens in the flock, will take a little time each day to fill out, and it will give the poultryman much needed information in order to know how to operate his business more successfully. Given these figures, it will be a simple matter to determine the cost of production and the profit per bird above the food consumed. A little notebook can be kept in which can be recorded records of sickness and death. Simple records can be evolved which will maintain in permanent form the results of hatching and brooding operations, and also a record of matings which are made for breeding purposes. These points are nothing but what any business man should want to know and would be expected to know in order to succeed in this present day of competition.

Keep a Simple Record of Money Transactions. A careful account should be kept of all money transactions. This account should be simple, but should show where every penny goes and from where every penny comes. By knowing the details of money exchange, the poultryman can plan each year to increase the parts of his work which are proving the most profitable, and curtail or decrease the less profitable portions.

The simple method of single entry which requires the debit and credit column is good, but it does not show at a glance for what money was spent or for what it was received. A modification of this method which takes no more time to keep, but which shows more detailed results, is known as the column system of single entry bookkeeping. The debit and credit sides of the book are subdivided into a number of columns, the credit or incoming side of the ledger being divided into such spaces as eggs, meat, live poultry, table, etc., the outgoing money side, or the debit side, can be ruled into such columns as feed, labor, supplies, etc. On adding up each of these columns, the totals will tell at a glance which products have returned the greatest revenue and which items of expenditure have required the greatest outlay of capital. Such an account book can be balanced at least once a month, and more often if so desired. The method is simple, yet it shows the poultryman at a glance the financial condition of his work. Plate No. 17 shows a good ruling for such a column system method.

Developing a System and Efficient Method of Carrying on of the Routine Work. The poultry business, being essentially an occupation of minor details, requires considerable thought and planning in order to insure that these details are accomplished each day with the same degree of care and completeness. This means that a definite plan for the doing of the chores must be worked out. The plan adopted requiring the least amount of unnecessary steps and doing the work each day with as much regularity and system as possible. The plan of work adopted should be written and posted conspicuously in the feed room or other suitable place in order that all affected by its provisions may know the routine. It will also serve as instructions and a guide to new help from time to time. A most excellent motto for the poultry raiser is

A TIME FOR EVERYTHING, WITH EVERYTHING DONE

IN ITS TIME.

CONCLUSIONS.

Efficiency, as measured by profit in dollars and cents, is the ultimate object of the poultry flock management. This efficiency can only be obtained by a thorough knowledge of the ten basic principles herein outlined and briefly discussed. A poultry business properly organized and managed by a man well experienced in the most efficient methods of production will surely be a remunerative and enjoyable occupation. This statement is true, both in its application to the operation of a specialized poultry farm or to the keeping of small flocks on farms or in suburban communities.

Know what your birds are doing and by modifying methods of production accordingly, be sure that your birds are keeping you rather than you keeping them.

PLATE NO. 17.

The column system of keeping accounts.

Incoming money or products sold.

Date.	To whom.	Eggs.	Meat.	Live Poultry.
10.	Henry Latham.	5 doz. at 30 cts. $1.50		
18.	Mrs. Johnson.		10 lbs. roaster at 25c—$2.50	
21	Harry Smith.	1 case eggs $7.60		
25.	C. H. Jones.			1 cockerel $2.00
	Total.			

Outgoing money or supplies purchased.

Date.	From whom.	Feed.	Labor.	Miscellaneous.
8.	People's Store.	100 lbs. wheat 200 lbs. corn $2.75		
10.	James Long for plowing yards.		$2.00	
26.	Sunnyside Poultry Farm.			Hatching eggs 30 for $2.00
	Total.			

Two sheets similar to the above can be ruled for each month in a year. One sheet each month can be used for incoming money, and the other for outgoing money as specified. They can be made long enough to accommodate the number of transactions for the month. At the end of the year these sheets can be bound together to make a permanent record.

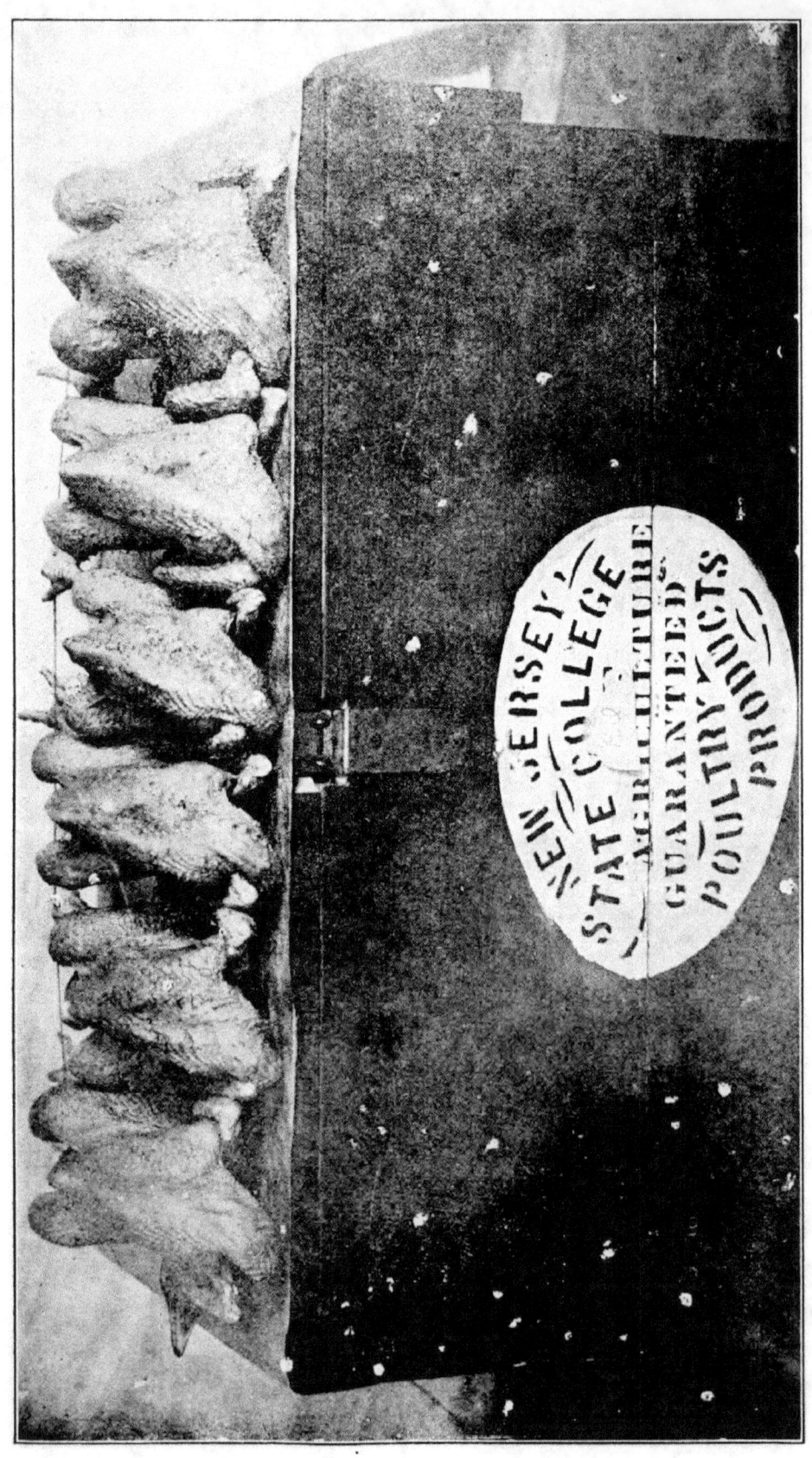

PLATE NO. 15.—Six prime large broilers of uniform size and grade. When packing dressed poultry for the market only birds which grade fairly uniform should be killed and shipped at one time.

PLATE NO. 14.—A very satisfactory way of shipping high quality eggs to market. The eggs packed in cartons and one-half of the case white and the other half brown. A heavy, durable case is used.

PLATE NO. 13.—An ideal range for the youngsters during the hot summer. Plenty of range, shade, green food and fresh air shelters are provided. Young peach trees are growing to make permanent shade in future years.

PLATE NO. 12.—A bunch of young huskies just out of the incubator. Is it any wonder that these little fellows must be given proper temperature and feed if they are to grow properly? No mother but a square box and oftentimes a careless man to care for them.

PLATE NO. 11.—A row of brooder houses in which are located colony brooder stoves. Each house has a capacity of 500 chicks. This is one of the most efficient and economical methods of brooding large numbers of chicks. It is a recent product of the poultryman's creative genius.

PLATE NO 10.—An incubator taken to pieces for cleaning and disinfecting, an operation which is essential at least once each year.

PLATE NO. 9.

(a) Belle of Jersey, a little White Leghorn Hen which laid during her pullet year 210 eggs and which consumed 40 times her own body weight to do this remarkable performance. She was the direct result of a systematic effort to breed for high egg production, and she has been the mother of many cockerels which have produced hundreds of heavy laying daughters.
(b) The 246 eggs which this hen laid during her first year's production.
(c) A comparison of the food she ate during this same period.

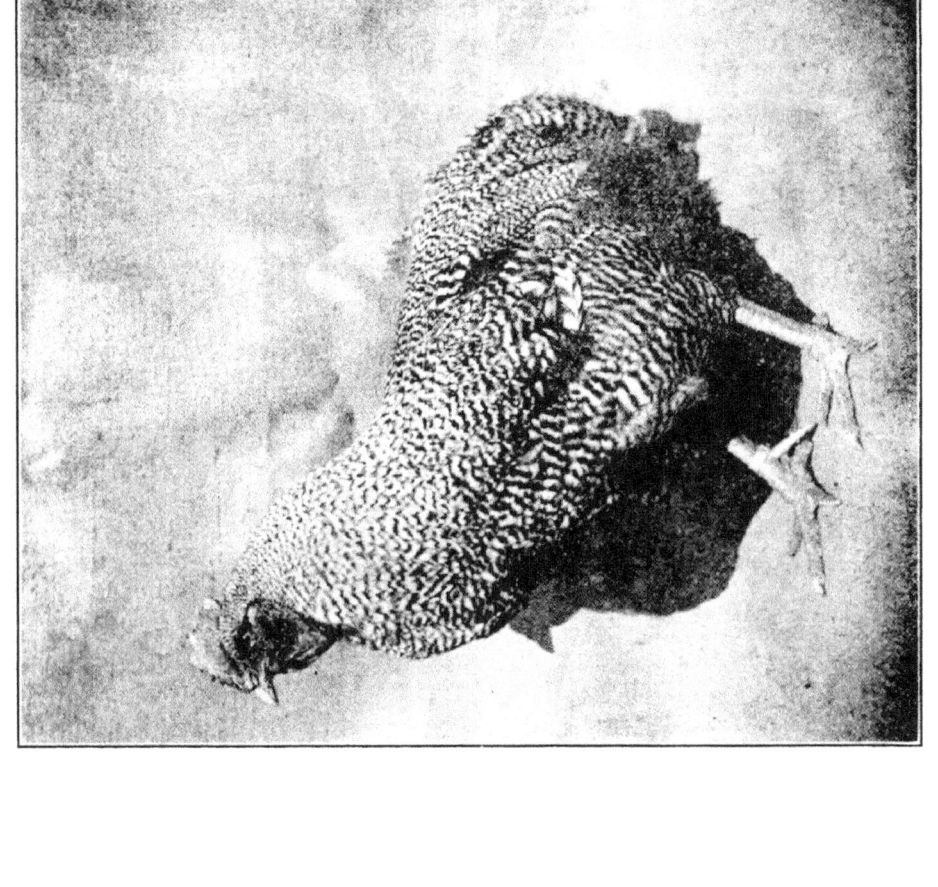

PLATE NO. 8.

(a) A typical example of vigor in the breeding male. The long body, moderately high tail, heavy shank, prominent eye and short beak are a few of the external indications of this vigor.
(b) A typical example of low vigor in a male bird. Note the low drooping carriage, long head and sunken eyes, which are always an indication of low vigor.

—Harvesting the Mangel Beets, showing a profitable practice of cutting the tops and feeding them green just before the roots are harvested.

PLATE NO. 5.—A large dry mash hopper built into the center of the house and so arranged that many birds can feed from it at one time. It is not wasteful of the mash, does not clog, and saves much labor. Also note the abundance of sunshine which covers much of the floor, making the house more cheerful.

PLATE NO. 4.—A view of a double section N. J. Multiple Unit Laying House. This type of construction is proving very popular and efficient under all conditions of climate and weather found in the State.

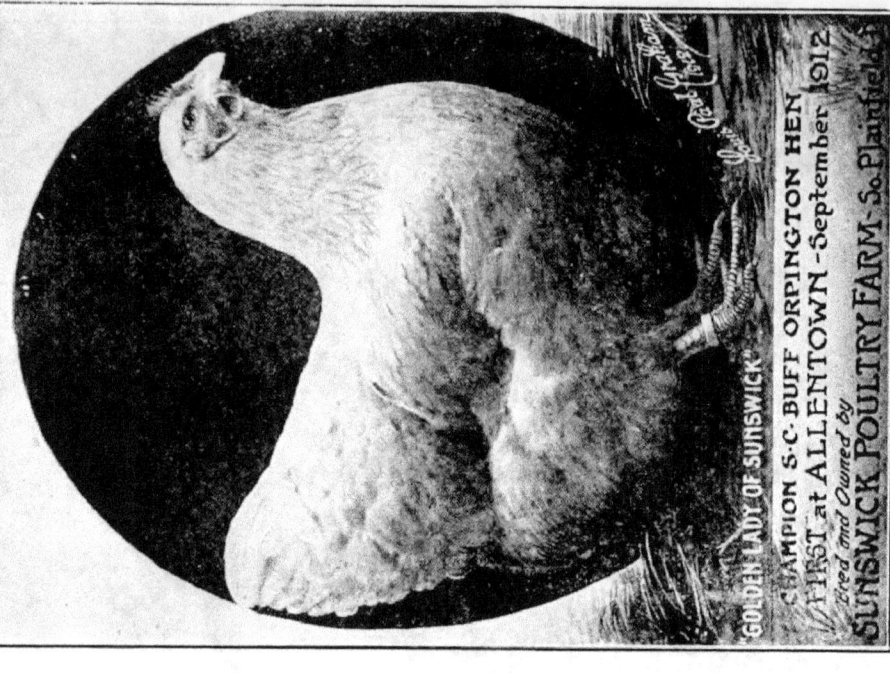

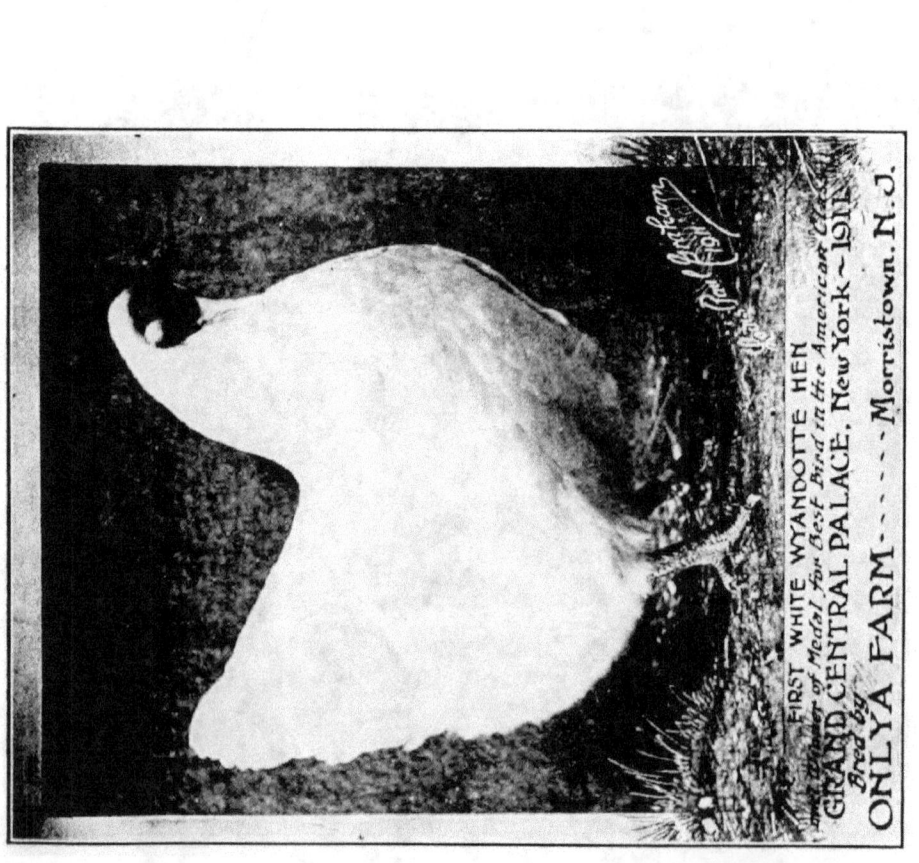

PLATE NO. 3.

(a) A beautiful specimen of the White Wyandotte variety. A type possessing wonderful dual purpose possibilities and very popular in the State.
(b) A remarkable Buff Orpington Hen. A representative of a typical English General Purpose breed which due to their capacity for egg and meat production have won their way to great favor at the hands of the American breeder.

www.ingramcontent.com/pod-product-compliance
Lightning Source LLC
Chambersburg PA
CBHW060000230526
45472CB00008B/1877